CRC Press
METHODS IN THE LIFE SCIENCES

Gerald D. Fasman - Advisory Editor
Brandeis University

D0103514

Series Overview

Methods in Biochemistry
John Hershey
Department of Biological Chemistry
University of California

Cellular and Molecular Neuropharmacology
Joan M. Lakoski
Department of Pharmacology
Penn State University

Research Methods for Inbred Laboratory Mice
John P. Sundberg
The Jackson Laboratory
Bar Harbor, Maine

Methods in Neuroscience
Sidney A. Simon
Miguel Nicolelis
Department of Neurobiology
Duke University

Methods in Pharmacology
John H. McNeill
Professor and Dean
Faculty of Pharmaceutical Science
The University of British Columbia

Methods in Signal Transduction
Joseph Eichberg, Jr.
Department of Biochemical and Biophysical Sciences
University of Houston

Methods in Toxicology
Edward J. Massaro
Senior Research Scientist
National Health and Environmental Effects Research Laboratory
Research Triangle Park, North Carolina

CRC Press
METHODS IN THE LIFE SCIENCES
— METHODS IN SIGNAL TRANSDUCTION

Joseph Eichberg, Jr., Editor

The CRC Press *Methods in the Life Sciences —Methods in Signal Transduction Series* provides the reader with state-of-the-art research methods that address the cellular and molecular mechanisms of the neuropharmacology of brain function in a clear and concise format.

Published Titles

Lipid Second Messengers, Suzanne G. Laychock and Ronald P. Rubin

LIPID SECOND MESSENGERS

Edited by

Suzanne G. Laychock

Department of Pharmacology and Toxicology,
School of Medicine and Biochemical Sciences,
State University of New York at Buffalo

and

Ronald P. Rubin

Department of Pharmacology and Toxicology,
School of Medicine and Biochemical Sciences,
State University of New York at Buffalo

CRC Press

Boca Raton   London   New York   Washington, D.C.

Library of Congress Cataloging-in-Publication Data

Catalog record is available from the Library of Congress.

Preface

The purpose of this volume is to provide detailed methodology for the analysis of a number of lipid signaling pathways and their products. It should serve as a valuable reference source for investigators wishing to utilize state-of-the-art methodologies for studying lipid second messengers. Emphasis focuses on procedures used to measure lipid-derived mediators, such as lysophospholipids, arachidonic acid, eicosanoids, anandamide, and ceramides, as well as the enzymes responsible for generating these messengers, such as phospholipases, prostaglandin endoperoxide synthases, and sphingomyelinase. In addition to providing new and important experimental tools to investigate novel lipid mediators in signal transduction, this volume includes important clues for successfully completing such experiments. Lipid mediators such as those covered in this volume not only play key roles in both signal transduction and cell-to-cell communication, but they will continue to be important factors in the ongoing development of novel and more effective therapeutic approaches.

The Editors

Suzanne Laychock, Ph.D., is Professor and Associate Chair of the Department of Pharmacology & Toxicology at the State University of New York at Buffalo, School of Medicine and Biomedical Sciences. She has been studying signal transduction mechanisms involving phospholipases, fatty acids, and eicosanoids since 1975. A major thrust of her work has been in defining the role of phospholipid hydrolysis and turnover, and prostaglandin synthesis in the control of endocrine cells. Her early pioneering work focused on the radioimmunoassay of prostaglandins and the characterization of calcium-dependent phospholipase A_2 in adrenocortical cells and isolated pancreatic islets. Her laboratory was also the first to demonstrate calcium regulation of phospholipase C–mediated hydrolysis of phosphoinositides. Dr. Laychock has published over 60 full publications and written three reviews and seven book chapters. She has served as a member of the Editorial Board of several prestigious journals.

Ronald P. Rubin, Ph.D., is Professor and Chair of the Department of Pharmacology & Toxicology at the State University of New York at Buffalo, School of Medicine and Biomedical Sciences. Dr. Rubin conducted pioneering work on the pivotal role of calcium in stimulus–response coupling in secretory cells. In addition, he has published widely on the role of lipid mediators in regulating the secretory process in a number of different cell types, including adrenocortical cells, pancreatic and parotid acinar cells, and neutrophils. His current research involves the signal transduction processes associated with the regulation of calcium signaling in parotid acinar cells. Dr. Rubin has published over 100 full publications, written several reviews, and authored two monographs. He also coedited a volume on *Calcium in Biological Systems*.

Contributors

Clary Clish
Center for Experimental Therapeutics
and Reperfusion Injury
Anesthesia Research
Brigham & Women's Hospital and
Harvard Medical School
Boston, Massachusetts

David DeWitt
Department of Biochemistry
Michigan State University
East Lansing, Michigan

David J. Fischer
Department of Physiology and
Biophysics
University of Tennessee, Memphis
Memphis, Tennessee

Louis Freysz
Laboratoire de Neurobiologie
Moleculaire des Interactions
Cellulaires
Centre de Neurochimie
Strasbourg, France

Andrea Giuffrida
The Neurosciences Institute
San Diego, California

Karsten Gronert
Center for Experimental Therapeutics
and Reperfusion Injury
Anesthesia Research
Brigham & Women's Hospital and
Harvard Medical School
Boston, Massachusetts

Zhong Guo
Department of Physiology and
Biophysics
University of Tennessee, Memphis
Memphis, Tennessee

Yusuf A. Hannun
Medical University of South Carolina
Charleston, South Carolina

Heather Hayter
Medical University of South Carolina
Charleston, South Carolina

Julian N. Kanfer
Department of Biochemistry and
Molecular Biology
Faculty of Medicine
University of Manitoba
Winnipeg, Manitoba, Canada

Suzanne Laychock
Department of Pharmacology and
   Toxicology
School of Medicine and Biomedical
   Sciences
State University of New York at Buffalo
Buffalo, New York

Bruce D. Levy
Center for Experimental Therapeutics
   and Reperfusion Injury
Anesthesia Research
Brigham & Women's Hospital and
   Harvard Medical School
Boston, Massachusetts

Károly Liliom
Institute of Enzymology
Biological Research Center
Hungarian Academy of Sciences
Budapest, Hungary

Daniele Piomelli
The Neurosciences Institute
San Diego, California

Ronald P. Rubin
Department of Pharmacology and
   Toxicology
School of Medicine and Biomedical
   Sciences
State University of New York at Buffalo
Buffalo, New York

Charles N. Serhan
Center for Experimental Therapeutics
   and Reperfusion Injury
Anesthesia Research
Brigham & Women's Hospital and
   Harvard Medical School
Boston, Massachusetts

Gábor J. Tigyi
Department of Physiology and Biophysics
University of Tennessee, Memphis
Memphis, Tennessee

Contents

Chapter

Phospholipase D

Louis Freysz and Julian N. Kanfer

Contents

I.   Introduction

Phospholipase D (PLD) activity has been detected in all eukaryotic cells examined.[1] The activation of PLD has been demonstrated in a variety of cells stimulated by agonists, including hormones, peptides, cytokines, neurotransmitters, growth factors, and nonphysiological stimuli, such as phorbol esters and calcium ionophores.[2,3]

0-8493-3383-0/99/$0.00+$.50
© 1999 by CRC Press LLC

1

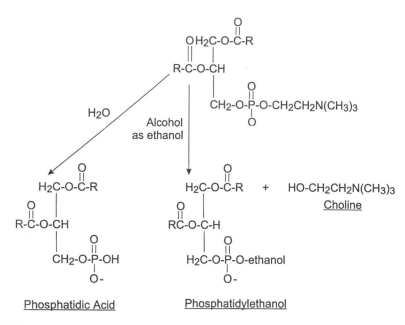

FIGURE 1
The reaction catalyzed by PLD using phosphatidylcholine as substrate.

Various observations provide evidence for the existence of multiple forms of PLDs having different subcellular localizations, distinct mechanisms of activation and substrate specificity, and different chromatographic properties.[4] The activation of PLD results in a transient production of phosphatidic acid. Because phosphatidic acid has been recognized as an important intracellular second messenger in multiple cellular responses,[5,6] PLD plays an essential role in the mechanisms of signal transduction and cell regulation.

The earliest report of PLD was of a phosphatidase C in carrot roots and cabbage leaves[7,8] employing phosphatidylcholine (PC) as the substrate. The rapid increase in recent publications about PLD was facilitated by the unusual property of PLD to catalyze a "transphosphatidylation" reaction. This reaction was originally described for cabbage PLD, which was found to produce several phosphatidyl alcohols in the presence of a spectrum of small primary alcohols, including glycerol, choline, and ethanolamine.[9,10] This property of PLD has been exploited to produce a variety of phospholipids and their analogs.[11] The fact that the transphosphatidylation property was also characteristic of rat brain PLD[12,13] explains why millimolar concentrations of ethanol are frequently found in PLD protocols.

Four distinct PLDs have been found in mammalian tissues. There are two separate PLDs that use PC and perhaps phosphatidylethanolamine as substrate. These two PLDs are operationally distinguished by the nature of the lipid required for activation. The earliest-described PLD requires an unsaturated fatty acid, as oleate, for activation. The more recently identified PLD requires a mixture of phosphatidylethanolamine and phosphatidylinositol bisphosphate in the presence of guanosine 5'-O-(3-thiotriphosphate), (GTPγS) for activation. A distinct lyso-PLD has been described, but not well studied. The other mammalian PLD that is specific for glycosylphosphatidylinositol provides the "anchor" for several enzymes and receptors at the external surface of the cell membrane. There are also numerous plant PLDs that have been described.

The estimation of PLD activity is based upon the measurement of an expected product. The reaction catalyzed by PLD is shown in Figure 1, using PC as substrate. Phosphatidic acid is the product obtained in the presence of water due to the hydrolytic property of the enzyme. In contrast, in the presence of methanol, ethanol, propranol, or butanol, the product is the corresponding phospholipid as phosphatidylethanol. The measurement of phosphatidic acid, phosphatidylethanol, or choline formation is employed as an index of PLD activity. There have been a variety of procedures employed to measure these products.

This chapter will provide the most frequently used methods for assessment of PLD activity. The *in vitro* assays use cell-free extracts and an exogenous substrate, usually PC. The intact cell assays use an undefined endogenous substrate and the PLD activity characteristic of the particular cell. There are similar techniques for product measurement for both the *in vitro* and intact cell studies.

II.   Protocol

A.   In Vitro Assays of PLD Activity

1.   Oleate-dependent PLD assay

The protocol described below employs radioactive PC as substrate and is similar to that previously reported.[14]

Procedure

---

1.   Prepare fresh micelles of substrate in sufficient quantities for each experiment to contain 12.5 mM fatty acid–labeled PC and 25 mM sodium oleate. Several forms are commercially available, but the dipalmitoyl species is a poor substrate. The substrate is composed of a mixture of egg yolk or beef brain PC (9.38 mg) and radioactive PC to yield a specific activity of 3000 to 5000 dpm/nmol; the mixture is placed in a glass tube and the solvent removed under a stream of N_2. A 1-ml aliquot of sodium oleate (25 mM) is added and the contents sonicated with a stick-type tip for 5 to 15 min with the tube in an ice bath. The final micellar solution should be clear and is used as the substrate.

2.   Be sure the final concentrations of the reagents in the incubation are

2.5 mM labeled PC

5 mM sodium oleate

50 mM β-dimethylglutaric acid buffer

10 mM ethylenediaminetetraacetic acid (EDTA)

25 mM NaF (to inhibit the phosphatidic acid phosphatase ubiquitous in mammalian cells)

0.3 M ethanol

and the enzyme source in a total volume of 120 µl.

3. Incubate the tubes for 60 min at 37°C, and then terminate the reaction by the addition of 3 ml chloroform/methanol (2:1) containing 15 µl each of the phosphatidic acid and the phosphatidylethanol standards.

4. Subject the samples to the classical Folch procedure for lipid extraction.[15] Take the final chloroform phase to dryness under an N_2 stream and dissolve in 50 to 100 µl chloroform/methanol (2:1).

5. Transfer the lipid extract to a thin-layer chromatography (TLC) plate of silica gel G. Separate the lipids in each sample using the upper organic phase from a mixture of ethylacetate/isooctane/glacial acetic acid/water (130:20:30:100) as a solvent. Dry the developed plates and expose them to iodine vapor. Collect the areas corresponding to the phosphatidic acid standard, the phosphatidylethanol standard, the unreacted substrate at the origin, and any diglyceride produced at the solvent front and quantitate the radioactivity with Scintiverse II.

Variations

1. In general, particulate PLD is not stimulated by divalent cations. However, 1 to 5 µM Ca^{2+} stimulates partially purified PLD from rat brain,[16] and 1 to 5 µM Ca^{2+} or Mg^{2+} stimulates a PLD purified from pig lung.[17] The presence of 1 mM Mg^{2+} and to a lesser extent 0.25 mM Ca^{2+} stimulates a neutral pH PLD of synaptic membranes.[18]

2. The incorporation of a radioactive alcohol into its corresponding phospholipid has been used to measure phospholipase D activity. This approach eliminates the need for a radioactive PC as substrate so that only the nonradioactive phospholipid substrate is required. The formation of phosphatidyl [3H]glycerol in the presence of [3H]glycerol,[19] the formation of phosphatidyl [14C]ethanol in the presence of [14C]ethanol,[20] and the formation of phosphatidyl [3H]butanol in the presence of [3H]butanol[21] can be used to estimate PLD activity.

3. The release of [3H]choline from dipalmitoyl-phosphatidyl [3H]choline has also been employed for the measurement of PLD activity.[17,22] The assay with this substrate is based upon the formation of a water-soluble, rather than a lipid-soluble product. There is a potential problem associated with the use of a choline-labeled substrate. It is possible that with crude homogenates the detection of choline could be the result of the sequential action of a phospholipase C, which would release a phosphocholine and a phosphatase that would hydrolyze phosphocholine to release free choline.

2.   Small-molecular-weight G protein–dependent PLD assay

There is present in mammalian tissues a form of PLD that is activated by small-molecular-weight GTP-binding proteins (G proteins). The *in vitro* assay of partially purified forms of this PLD requires the inclusion of one of these G proteins. The presence of the cytosolic components of a tissue homogenate usually is sufficient for assay of cell-free preparations. GTPγS, a nonhydrolyzable GTP analog, is a necessary constituent of the assays. Phosphatidylinositol 4,5 bisphosphate (PIP₂) indirectly facilitates the binding of GTP to these G proteins and is also a component of the assay mixture. The protocol is similar to that previously published.[23]

Procedure

1. Prepare fresh micelles in sufficient quantity for each experiment, to contain 1 mM phosphatidylethanolamine/150 μM PIP$_2$/100 μM phosphatidyl [^{3}H]choline (specific activity of 100,000 dpm/nmol). Remove solvents with an N$_2$ stream and sonicate the residue in a solution composed of 100 mM HEPES buffer, pH 7.5; 6 mM ethylene-glycol-bis (β-aminoethyl ether)-tetraacetic acid (EGTA); 160 mM KCl; 2 mM dithio-threitol (DTT). Sonicate the sample with a stick-type tip for 30 s to produce a clear micelle solution and use this as the substrate.

2. Be sure the final reaction mixture (100 μl final volume) is composed of

    33 mM HEPES pH 7.5

    1.9 mM EGTA

    0.66 mM DTT

    53 mM KCl

    3 mM MgCl$_2$

    2 mM CaCl$_2$

    100 μM phosphatidylethanolamine

    15 μM PIP$_2$

    10 μM PC

    10 μM GTPγS

    100 μM ATP

    and an enzyme source and a small-molecular-weight G-protein source usually present in the tissue extract being assayed.

    Incubate the tubes for 60 min at 37°C; then transfer them to an ice bucket and terminate the reaction by the addition of 100 μl of 100 mg/ml bovine serum albumin (BSA) followed by 100 μl of 10% trichloroacetic acid (TCA). Mix the samples well and centrifuge at maximum speed in a microfuge in the cold.

3. If desired, use the supernatant containing the [^{3}H]choline released directly for liquid scintillation counting. Alternatively, reduce the TCA content of the supernatant by extraction three times with 1 ml diethylether. To the extracted aqueous phase, add 5 μl of 0.8 N NaOH to partially neutralize the solution. A 50-μl aliquot can then be subjected to Fonnum partitioning.[24] This procedure employs an organic solvent soluble chelator for extraction of choline and acetylcholine.

Variations

1. Fatty acid–labeled PC can replace the choline-labeled substrate and in the presence of 0.3 M ethanol the formation of phosphatidylethanol measured. It has been reported that the presence of 1.6 M ammonium sulfate in these incubations causes a 72-fold stimulation of the enzyme activity.[25]

3.   Nonmammalian PLD assay

There are a variety of PLDs isolated from plants or microorganisms. In general, they require a modest quantity of calcium and a detergent to "solubilize" the substrate, most frequently PC, and a buffer at an acidic pH. A representative description for the PLD from peanut seeds[26] or cabbage[11] illustrates the principal.

B. Assay of PLD Activity with Intact Cells

The *in vitro* assays of PLD described in Section II.A utilize a defined exogenous substrate. In contrast, the substrate for PLD assays with intact cells is endogenous and usually undefined. The general approach for these assays is to label intact cells with a radioactive precursor. Usually these are general precursors for phospholipids rather than for a specific phospholipid. The common precursors employed are labeled fatty acids, most frequently [3H]myristic acid, [3H]glycerol, or inorganic $^{32}PO_4$. [3H]Choline would be expected specifically to label principally PC but could also label lyso-PC and sphingomyelin. There would also be intracellular pools of [3H]choline, phosphoryl[3H]choline and perhaps glycerophosphoryl [3H]choline. The appearance of [3H]choline is used as the index of PLD activity. However, it is possible that the [3H]choline measured could have arisen from the other labeled pools. This is a serious limitation in the use of [3H]choline as a labeled precursor. In addition, the commercially available cell culture medium contains phosphate salts and choline as constituents, which would dilute the corresponding radioactive precursors by the nonradioactive compound.

Some of these limitations can be avoided by employing a labeled alkyl-lyso-phospholipid. These compounds are rapidly taken up by most cells and rapidly acetylated and converted into a phospholipid. The acyl-lysophospholipids but not the alkyl-lysophospholipid may be hydrolyzed by a lysophospholipase and merely provide a labeled fatty acid precursor. The conversion of an alkyl-lyso-PC will provide an alkyl-acyl-PC as the only labeled pool and a well-defined substrate for PLD.

The measurement of PLD activity in intact cells is dependent upon an increase from a finite baseline of an expected product. As described in Section II.A for the *in vitro* assay, the presence of a defined activator such as sodium oleate or a cocktail of GTPγS and a small-molecular-weight G protein is essential. The activation of PLD present in intact cells requires the interaction of an agonist with specific receptors on the external surface of the cell. This coupling of agonist binding to the cell triggering the stimulation of PLD is dependent upon the receptor profile of the particular cell type.

There are thousands of publications describing the activation of PLD by a particular agonist in a particular cell type. Some general guidelines and limited references to publications using the particular radioactive precursor will be provided. This does not represent any recommendation but merely serves as an illustration of the approach. The protocols employing the precursors expected to be incorporated into the "phosphatidyl" portion of a phospholipid as labeled fatty acid, glycerol or $^{32}PO_4$ also include an alcohol such as ethanol. The extraction of cellular lipids, their TLC separations, and the measurements of radioactivity present in phosphatidic acid and phosphatidylethanol are described in Section II.A.

A detailed description for the assay of PLD activation using [3H]palmitate and [3H]choline prelabeled fibroblast cell cultures has appeared[27] and the reader is referred to it. This technique can be adapted for other cell cultures labeled with different phospholipid precursors.

Procedure

1. Label cells with $^{32}PO_4$ by employing 50 to 200 µCi/ml of $^{32}PO_4$ for 1 h or overnight[28]; glycerol by employing 30 to 50 µCi/ml [³H]glycerol for several hours or overnight[29]; fatty acids by employing 10 µCi/ml [³H]palmitic acid for 24 h[30] or 5 µCi/ml [³H]myristic acid for 18 h.[31] Carry out the labeling in the growth medium specified for the particular cell type. The challenge with an agonist is often conducted in a simpler medium.

2. Label with lysophospholipid by employing 5 µCi [³H]alkyl-lyso-PC for 30 min to 2 h in a buffered salt solution rather than with growth medium.[32]

3. Label with choline by employing 0.3 to 1.5 µCi/ml [³H]choline for 24 to 48 h in a growth medium. Ethanol is usually absent from the protocols in the experiments with choline-prelabeled cells.[33] The quantitation of [³H]choline release has been estimated by various methods including those mentioned in Section II.A. Radioactivity released in the incubation mixture has been determined directly, or in the aqueous phase after a Folch extraction.[15] These results will not distinguish between the several possible [³H]choline-containing water-soluble metabolites. Free [³H]choline in the incubation media also has been determined by the Fonnum technique.[24] TLC of the water-soluble substances with 0.5% $NaCl/CH_3OH/28\%$ NH_4OH (50:50:1) as solvent has been used to distinguish choline, phosphocholine, and glycerophosphocholine.[33] Separation of choline, phosphocholine, and glycerophosphorylcholine can be accomplished using Dowex-50 H^+ ion exchange column chromatography.[34] A similar methodology has been used for measurement of [³H]ethanolamine containing water-soluble metabolites from [³H]ethanolamine-prelabeled cells.[35]

Variation

1. PLD activation of intact cells has been detected using a radioactive alcohol rather than by labeling a pool of phospholipase substrates. These incubations were conducted in the presence of 62 µCi/ml of [¹⁴C]ethanol[36] or 200 µCi [³H]butanol[37] and the radioactivity appearing in the corresponding phosphatidyl alcohol measured.

III. Alternative Assay Methods for PLD Activity

A. Fluorescent Assay

Fluorescent derivatives of PC have been employed as substrates. These are bulky derivatives of fatty acid esterified at the C-2 position of the PC. These include BIODIPY, 4,4 difluoro-5,7, dimethyl-4-bora-3a,4a diaza-*s* indacene; nitrobenzoxa-diazote (NBD),2-(12-(7 nitrobenz-2-oxa-1,3-diazol-4-yl) amino, and HP, 1-pyrende-canoyl derivatives. An example is the use of BIODIPY undecylphosphatidyl choline as substrate in the presence or absence of an alcohol. The lipids are extracted at the termination of the experiments and separated by TLC in a manner similar to that described in Section II.A. The TLC plates are analyzed with a fluorescence-scanning densitometer with excitation at 507 nM and emission at 514 nM. This quantitates phosphatidic acid, phosphatidylethanol, or phosphatidylbutanol formation, which are indexes of PLD activity.[38] In the absence of a TLC scanner, the spots can be visualized with an ultraviolet lamp, removed from the plate, extracted with ethanol, and the intensity measured in a fluorescence spectrophotometer set at 507 nM for the excitation wavelength and 514 nM for the emission wavelength.[39]

B.   Spectrophotometric Assays

Spectrophotometric assays of *p*-nitrophenyl analogs as substrates have been suggested as useful for PLD assays. *p*-Nitrophenyl-phosphocholine was incubated with plant PLD in the presence of acid phosphatase. The hydrolysis would be expected to produce free choline and *p*-nitro phenylphosphate. The acid phosphatase would cleave the latter product to free *p*-nitrophenol, which can be quantitated by absorbance at 400 nM.[40] Phosphatidyl *p*-nitrophenol was prepared as a PLD substrate[41] and utilized in the assay for the purification of a bacterial PLD.[42]

C.   Choline Oxidase–Based Assays

1.   The quantitation of free choline has been based upon the use of the enzyme choline oxidase, which catalyzes the following reaction:

$$Choline + H_2O + 2O_2 \rightarrow betaine + H_2O_2$$

The hydrogen peroxide is hydrolyzed by a peroxidase and in the presence of a suitable dye measured spectrophotometrically. This principle has led to an high performance liquid chromatography (HPLC) procedure. Choline oxidase and peroxidase are bound to a support to provide an enzyme reactor combined with an HPLC separation system and an electrochemical detector.[43] This approach has been exploited to measure the activity of PLD.

2.   Rather than a chromogenic detection, one based upon light release has been developed according to the sequence[44]:

$$Choline + 2O_2 + H_2O \rightarrow betaine + 2H_2O_2 \text{ (by choline oxidase)}$$

$$Luminol + 2H_2O_2 \rightarrow aminophalate + 4H_2O + N_2 + light \text{ (by microperoxidase)}$$

The usual incubations mixtures (200 μl) for PLD contain in addition 0.3 to 1 U of choline oxidase (Alcaligenes). At the end of this reaction the tubes receive 50 μmol luminol in a $NaCO_3/NaHCO_3$ buffer + 20 μU of microperoxidase in 10 nM Tris buffer pH 7.4. The chemiluminescence measured in a luminometer will detect 10 pmol of free choline. An automated flow injection analysis system for monitoring PLD has been developed using luminol chemoluminescence.[45]

3.   A coupled spectrophotometric method for kinetic estimates of PLD activity has been described.[46] This depends upon the peroxidase-catalyzed hydrolysis of H_2O_2 coupled

to 4-aminoantipyrene (AAP) and 2-hydroxy-3,5 dichlorobenzene sulfonate (HDCBS). The red-colored chromagen formation is monitored at 510 nM in a spectrophotometer. The reaction mixture contains 300 µl of 60 to 70 µM PC (dissolved in a mixture of 0.9 g/l sodium dodecyl sulfate (SDS) and 2.0 g/l Triton X-100); 10 µl of 100 mM Tris, pH 8; 500 µl of 50 mM Tris/18 mM HDCBS/4.8 nM AAP/10 U peroxidase/5 U of choline oxidase; and 50 µl of 0.18 M calcium chloride. The addition of PLD initiates the reaction.

D. Choline Kinase–Based Assays

1. There are descriptions of choline measurements based upon the phosphorylation by choline kinase, which catalyzes the following reaction:

$$\text{Choline} + \text{ATP} \rightarrow \text{phosphocholine} + \text{ADP}$$

The ADP produced can be measured by coupling with pyruvate kinase and lactate dehydrogenase to measure PLD activity[47] according to the following reaction:

$$\text{ADP} + \text{phosphoenolpyruvate} \rightarrow \text{ATP} + \text{pyruvate (pyruvate kinase)}$$

$$\text{pyruvate} + \text{NADH} + \text{H}^+ \rightarrow \text{lactate} + \text{NAD (lactic dehydrogenase)}$$

The oxidation of NADH to NAD is measured spectrophotometrically at 340 nM.

2. Choline kinase provides the basis for another method to quantitate choline liberated as a result of PLD activity. The basis for this method is the use of [^{32}P]ATP of known specific activity to produce [^{32}P]phosphocholine. This [^{32}P]phosphocholine is separated from the remaining [^{32}P]ATP by passage through an anion exchange resin.[48,49]

E. Novel PLD Assays

1. Several novel methods have been described for PLD assays. ^{1}HNMR spectroscopy has been used to measure the hydrolysis of PC liposomes by a *Streptomyces* PLD. The ability to distinguish the chemical shifts of the protons of the methyl groups of free choline from the chemical shifts of the choline methyl protons of PC is the basis of this determination.[50] The increase of free choline upon the addition of calcium ions to red blood lysates measured by ^{1}H spin-echo nuclear magnetic resonance (NMR) may be an index of PLD hydrolysis of PC.[51] PC is a zwitterion. Hydrolysis produces two independent mobile ionic species. This change can be monitored by solution electrical conductance and has been employed to assay PLD activity.[52]

IV. Assays of Lysophospholipase D Activity

Lyso-PLD activity was discovered in rat brain microsomes[53] but has not been purified. The properties of this enzyme have been established utilizing microsomes from several mammalian tissues. This enzyme will use 1-0-alkyl but not 1-0-acyl lysoglyc-erophosphocholine or lyso-glycerophosphoethanolamine as substrates.[54] Based upon the specificity for the 1-0-alkyl species, lyso-PLD may have a role in the metabolism of platelet-activating factor (PAF).

Microsomal preparations treated with EDTA require either Ca^{2+} or Mg^{2+} as a component of the incubation mixture. Incubations with untreated microsomes do not require the presence of either cation. The basis for the assay of lyso-PLD is the formation of a 1-0-alkyl lysoglycero-3 phosphate or 1-0-alkylglycerol. Details of an assay system have been published.[55] Microsomes are incubated with 1-0[³H] alkyl-2-lyso-*sn*-glycero-3-phosphocholine, 0.1 *M* Tris buffer pH 7.2 or 8.4 with or without $CaCl_2$ or $MgCl_2$. Lipids are extracted at the termination of the incubation and separated by TLC. The major product is usually 1-0-alkyl glycerol because of the ubiquitous, active lysophosphatidic acid phosphatase present in the microsomal preparation.

V. Assays of Glycosyl-Phosphatidylinositol–Specific PLD Activity Reagents Needed

There are a variety of proteins that are tethered to the external surface of cells through a glycosylphosphatidylinositol (GPI) anchor. The phosphatidylinositol portion is embedded in the membrane. A variety of binding proteins, enzymes, and some receptors are linked through a phosphoethanolamine and a series of mannosyl and an *N*-acetylglucosamine residue to the inositol moiety of the phosphatidylinositol.[56] Mammalian plasma and serum have a specific PLD that hydrolyzes the GPI. This releases the protein bound to the glycosyl residues, and the phosphatidic acid is retained in the cellular membrane. The assays devised for measuring GPI–PLD are more complicated than those developed for the other PLDs. The initial task is to obtain a suitable substrate for the enzyme assay. There are two different substrates that have been utilized.

Alkaline phosphatase with an intact GPI anchor isolated from calf intestine or human placenta provides one of the substrates. The basis for the assay is the conversion of a membrane-bound or hydrophobic form of alkaline phosphatase to a hydrophilic form. The assay conditions must also provide an environment that does not denature the alkaline phosphatase activity. The basis for the measurement of the GPI–PLD is the hydrophilic alkaline phosphatase activity. The details for the substrate preparations and the two-step assays have been published.[57]

Procedure

---

1. Incubate the GPI-linked alkaline phosphatase with the GPI–PLD source in a 200-mM Trismaleate buffer (pH 7.0) containing 1% Nonidet P40 in a volume of 200 µl for 30 min at 37°C.

2. Dilute samples into 800 µl of a solution containing ice-cold 150 mM NaCl/0.1 mM $MgCl_2$/10 µM zinc acetate/10 mM M HEPES pH 7.0, and take an aliquot directly for the measurement of alkaline phosphatase activity. This value represents the total amount used as the substrate for the GPI–PLD.

3. Mix another aliquot of the diluted sample with Triton X-114. Incubate this solution at 37°C for 10 min and centrifuge at room temperature. Take the alkaline phosphatase present in the upper hydrophilic phase as an index of GPI–PLD activity.

Variations

1. To prepare a radioactive substrate, the parasitic protozoan *Trypanosoma brucei* MIT at 118 or 117 is harvested from the blood of rats previously infected with this organism. The trypanosomes are exposed in culture to 2 mCi [^{3}H]myristic acid and after a suitable period of time the GPI-linked variant surface glycoprotein is isolated. This provides the substrate for the assays of GPI–PLD by measurement of [^{3}H]phosphatidic acid formation.[57]

2. An assay system avoiding the use of Triton X-114 with the substitution of a small octyl Sepharose column has been described. The alkaline phosphatase–linked GPI is used as the substrate. The product of GPI–PLD activity is recovered from the column with a solution of 0.025% Triton X-100 in diethanolamine buffer (pH 9.8). The alkaline phosphatase activity is used as an index of GPI–PLD activity.[58]

Reagents Needed _____

*β-Dimethylglutaric Acid Buffer, pH 6.5, 500 m*M

*EDTA, pH 6.5, 200 m*M

*NaF, 250 m*M

*Na Oleate, 25 m*M *(freshly prepared)*

*Ethanol, 8*M

Phosphatidic Acid Solution
   1 mg phosphatidic acid/ml chloroform/methanol (2:1)

Phosphatidylethanol Solution
   1 mg phosphatidylethanol/ml chloroform/methanol (2:1)

HEPES Buffer
   100 mM HEPES, pH 7.5
   6 mM EGTA
   160 mM KCl
   2 mM DTT
   9 mM MgCl$_2$
   6 mM CaCl$_2$

GTPγS

        0.2 mM GTPγS
        1 mM ATP

VI. Discussion

Several PLDs are detectable in mammalian tissues, but their possible individual physiological functions are based largely upon circumstantial observations. The most frequently proposed physiological role is the production of the biologically active substances lysophosphatidic acid, phosphatidic acid, and diglyceride. Although PLD appears to require an activator both with intact and broken/permeabilized cell preparations, the mechanisms involved in activation are unclear because the activating substances do not appear to be recognized cofactors required for the catalytic activity.

The assay procedures described for the intact and broken/permeabilized cell determinations of PLD activity should be regarded as complementary. The fact that a Protein Kinase C (PKC) activator, such as a phorbol ester, activates PLD of intact cells suggests that PKC is involved. However, this does not *a priori* mean that PLD phosphorylation by PKC is the mechanism of activation. In fact, the ability of purified preparations of PKC to activate partially purified or membrane-bound PLD by an ATP-independent mechanism suggests that phosphorylation is not required.[1] A PLD activation by GTPγS and an inhibition by GDPβS with permeabilized cells would suggest involvement of small-molecular-weight G proteins[2] or heterotrimeric G proteins. The mechanisms involved in PLD activation can be clarified by conducting the appropriate studies in intact- and broken-cell preparations.

References

1. **Exton, J. H.,** Phospholipase D: enzymology, mechanisms of regulation and function, *Physiol. Rev.*, 77, 303, 1997.
2. **Billah, M. M.,** Phospholipase D and cell signaling, *Curr. Opin. Immunol.*, 5, 114, 1993.
3. **Exton, J. H.,** Phosphatidylcholine breakdown and signal transduction, *Biochem. Biophys. Acta*, 1212, 26, 1994.
4. **Liscovitch, M. and Chalifa-Caspi, V.,** Enzymology of mammalian phospholipase D: *in vitro* studies, *Chem. Phys. Lipids*, 80, 37, 1996.
5. **Kroll, M. H., Savoico, G. B., and Schafer, A. I.,** Second messenger function of phosphatidic acid in platelet activation, *J. Cell. Physiol.*, 139, 558, 1989.
6. **Lang, D., Malviya, A. N., Hubsch, A., Kanfer, J. N., and Freysz, L.,** Phosphatidic acid activation of protein kinase C in LA-N-1 neuroblastoma cells, *Neurosci. Lett.*, 201, 199, 1995.
7. **Hanahan, D. J. and Chaikoff, I. L.,** A new phospholipid splitting enzyme specific for the ester linkage between the nitrogenous base and the phosphoric acid group, *J. Biol. Chem.*, 169, 699, 1947.

8. **Hanahan, D. J. and Chaikoff, I. L.,** On the nature of the phosphorus-containing lipids of cabbage leaves and their relation to a phospholipide-splitting enzyme contained in these leaves, *J. Biol. Chem.*, 172, 191, 1948.

9. **Dawson, R. M. C.,** The formation of phosphatidylglycerol and other phospholipids by the transferase activity of phospholipase D, *Biochem. J.*, 102, 205, 1967.

10. **Yang S. F., Freer, S., and Benson, G. A.,** Transphosphatidylation by phospholipase D, *J. Biol. Chem.*, 242, 477, 1967.

11. **Eibl, H. D. and Kovatchev, S.,** Preparation of phospholipids and their analogs by phospholipase D, *Methods Enzymol.*, 72, 632, 1981.

12. **Gustavson, G. and Alling, C.,** Formation of phosphatidylethanol in rat brain by phospholipase D, *Biochem. Biophys. Res. Commun.*, 142, 958, 1987.

13. **Kobayashi, M. and Kanfer, J. N.,** Phosphatidylethanol formation via transphosphatidylation by rat brain synaptosomes, *J. Neurochem.*, 48, 1597, 1987.

14. **Kobayashi, M. and Kanfer, J. N.,** Solubilization and purification of rat tissue phospholipase D, *Methods Enzymol.*, 197, 575, 1991.

15. **Folch-Pi, J., Lees, M., and Sloane-Stanley, G. M.,** A simplified method for the isolation and purification of total lipids from animal tissues, *J. Biol. Chem.*, 26, 497, 1957.

16. **Taki, T. and Kanfer, J. N.,** Partial purification and properties of rat brain phospholipase D, *J. Biol Chem.*, 254, 9761, 1979.

17. **Okamura, S. and Yamashita, S.,** Purification and characterization of a phosphatidylcholine phospholipase D from pig lung, *J. Biol. Chem.*, 264, 31207, 1994.

18. **Chalifa, V., Mohn, H., and Liscovitch, M.,** A neutral phospholipase D activity from rat brain synaptic membrane, *J. Biol. Chem.*, 265, 17512, 1990.

19. **Chalifour, R. J., Taki, T., and Kanfer, J. N.,** Phosphatidylglycerol formation via transphosphatidylation by rat brain extracts, *Can. J. Biochem.*, 58, 1189, 1980.

20. **Kanfer, J. N. and Hattori, H.,** Mammalian phospholipase D and related activities, in *Enzymes of Lipid Metabolism II*, Freysz, L., Dreyfus, H., Massarelli, R., and Gatt, S., Eds., Plenum Press, New York, 1986, 665.

21. **Horwitz, J. and Davis, G. L.,** The substrate specificity of brain microsomal phospholipase D, *Biochem. J.*, 295, 793, 1993.

22. **Massenburg, D., Han, J. S., Liyanage, M., Patton, W. A., Rhee, S. G., Moss, J., and Vaughan M.,** Activation of rat brain phospholipase D by ADP-ribosylation factors, 5 and 6: separation of ADP ribosylation factor-dependent and oleate-dependent enzymes, *PNAS*, 91, 11718, 1994.

23. **Brown, H. A. and Sternweis, P. C.,** Stimulation of phospholipase D by ADP — ribosylation factor, *Methods Enzymol.*, 257, 313, 1995.

24. **Fonnum, F.,** A rapid radiochemical method for determination of choline actyltransferase, *J. Neurochem.*, 24, 407, 1975.

25. **Nakamura, S. J., Shimooku, K., Akisue, T., Jinnai, H., Hitomic, T., Kiyohara, J., Ogino, C., Yoshida, K., and Nishizuka, Y.,** Mammalian phospholipase D: activation by ammonium sulfate and nucleotides, *PNAS*, 92, 12319, 1995.

26. **Heller, M., Mozes, N., and Mals, E.,** Phospholipase D from peanut seeds, *Methods Enzymol.*, 35, 226, 1975.

27. **Wakelam, M. J. O., Hodgkin, M., and Martin, A.,** The measurement of phospholipase D linked signaling in cells, in *Methods Molecular Biology*, Vol. 41, *Signal Transduction Protocols*, Kendall, D. A. and Hill, S. J., Eds., Humana Press, Totowa, NY, 1995, 271.

28. **Chiang, T. M.,** Activation of phospholipase D in human platelets by collagen and thrombin and its relationship to platelet aggregation, *Biochem. Biophys. Acta*, 1224, 147, 1994.

29. **Fan, X. T., Sherwood, J. L., and Haslam, R. J.,** Stimulation of phospholipase D in rabbit platelets by nucleoside phosphates and by phosphocreatine: role of membrane-bound GDP, nucleoside diphosphate kinase and creatine kinase, *Biochem. J.*, 299, 701, 1994.

30. **Horwitz, J.,** Bradykinin activates a phospholipase D that hydrolyzes phosphatidylcholine in PC12 cells, *J. Neurochem.*, 56, 509, 1991.

31. **Singh, I. N., Massarelli, R., and Kanfer, J. N.,** Activation of phospholipases D and A by amphiphilic cations of cultured LA-N-2 cells is G-protein and protein kinase C independent, *J. Lipid Med.*, 7, 85, 1993.

32. **Geny, B. and Cockcroft, S.,** Synergistic activation of phospholipase D by protein kinase C and G-protein mediated pathways in streptolysin σ-permeabilized HL60 cells, *Biochem. J.*, 284, 531, 1992.

33. **Martenson, E. A., Goldstein, D., and Brown, J. H.,** Muscarinic receptor activation of phosphatidylcholine hydrolysis, *J. Biol. Chem.*, 264, 14748, 1989.

34. **Cook, S. J. and Wakelman, M. J. O.,** Analysis of water soluble products of phosphatidylcholine breakdown by ion exchange chromatography, *Biochem. J.*, 263, 581, 1989.

35. **Kiss, Z. and Crilly, K. D.,** Adenosine triphosphate (ATP) and other nucleotides stimulate the hydrolysis of phosphatidylethanolamine in intact fibroblasts, *Lipids*, 36, 777, 1991.

36. **Metz, S. G. and Dunlop, M.,** Production of phosphatidylethanol by phospholipase D phosphatidyltransferase in intact or dispersed pancreatic islets: evidence for the *in situ* metabolism of phosphatidylethanol, *Arch. Biochem. Biophys.*, 283, 417, 1990.

37. **Randall, R. W., Bonser, R. W., Thompson, N. T., and Garland, L. G.,** A novel and sensitive assay for phospholipase D in intact cells, *FEBS Lett.*, 264, 87, 1990.

38. **Ella, K. M., Meier, P., Bradshaw, C. D., Huffman, K. M., Spivey, E. C., and Meier, K. E.,** A fluorescent assay for agonist-activated phospholipase D in mammalian cell extracts, *Anal. Biochem.*, 218, 136, 1994.

39. **Fujita, K. I., Murakami, M., Yamashita, F., Amemija, K., and Kudo, I.,** Phospholipase D is involved in cytosolic phospholipase A_2-dependant selective release of arachidonic acid by fMLP-simulated rat neutrophils, *FEBS Lett.*, 395, 293, 1996.

40. **Munishwa, N. G. and Wold, F.,** A convenient spectrophotometric assay for phospholipase D using *p*-nitro-phenylphosphocholine, *Lipids*, 15, 594, 1980.

41. **Arrigo, P. D., Piergianni, V., Scarcelli, D., and Servi, S.,** A spectrophometric assay for phospholipase D, *Anal. Chim. Acta*, 304, 249, 1995.

42. **Carrea, G., Arrigo, P. D., Piergianni, V., Roncaglio, S., Secundo, F., and Servi, S.,** Purification and properties of two phospholipase D from *Streptomyces* sp., *Biochem. Biophys. Acta*, 1255, 273, 1995.

43. **Koshimura, K., Miwa, S., Lee, K., Hayashi, Y., Hasegawa, H., Hamata, K., Fugiwara, M., Kimura, M., and Itokawa, Y.,** Effects of choline administration on *in vivo* release and biosynthesis of acetylcholine in rat striatum as studied *in vivo* brain microdialysis, *J. Neurochem.*, 54, 533, 1990.

44. **Hasegawa, Y., Kunou, E., Shindo, J., and Yuki, H.,** Determination of platelet activating factors by a chemiluminescence method and its application to stimulated guinea pig neutrophils, *Lipids,* 26, 1117, 1991.
45. **Becker, M., Spohn, U., and Ulbrich-Holfmann, R.,** Detection and characterization of phospholipase D by flow injection analyses, *Anal. Biochem.,* 244, 55, 1997.
46. **Takrama, J. F. and Taylor, K. E.,** A continuous spectrophotometric method monitoring phospholipase D catalyzed reactions of physiological substrates, *J. Biochem. Biophys. Methods,* 23, 217, 1991.
47. **Carman, G. M., Fischl, A. S., Dougherty, M., and Malcker, G.,** A spectrophotometric method for the assay of phospholipase D activity, *Anal. Biochem.,* 110, 73, 1981.
48. **Muma, N. A. and Rowell, P. P.,** A sensitive and specific radioenzymatic assay for the simultaneous determination of choline and phosphatidylcholine, *J. Neurosci. Methods,* 12, 249, 1985.
49. **Murray, J. J., Derik, T. T., Truett, A. P., and Kennerly, D. A.,** Isolation and enzymatic assay of choline and phosphocholine present in cell extracts with pmole sensitivity, *Biochem. J.,* 270, 63, 1990.
50. **Dorovska-Taron, V., Wick, C., and Walde, P.,** A ^{1}H nuclear magnetic resonance method for investigating the phospholipase D-catalyzed hydrolysis of phosphatidylcholine in lysosomes, *Anal. Biochem.,* 240, 37, 1996.
51. **Selle, H., Chapman, B. E., and Kuckel, P. W.,** Release of choline by phospholipase D and a related phosphoric diester hydrolase in human erythrocytes, *Biochem. J.,* 284, 61, 1992.
52. **Mezna, M. and Lawrence, A. J.,** Conductimetric assays for the hydrolase and transferase activity of phospholipase D enzyme, *Anal. Biochem.,* 218, 370, 1994.
53. **Wykle, R. and Schremmer, J. M.,** A lysophospholipase D pathway in the metabolism of ether-linked lipids in brain microsomes, *J. Biol. Chem.,* 249, 1742, 1974.
54. **Wykle, R. L., Kramer, W. F., and Schremmer, J. M.,** Specificity of lysophospholipase D, *Biochem. Biophys. Acta,* 619, 58, 1980.
55. **Wykle, R. L. and Strum, J. C.,** Lysophospholipase D, *Methods Enzymol.,* 197, 583, 1991.
56. **Low, M. G.,** The glycosyl-phosphatidylinositol anchor of membrane proteins, *Biochem. Biophys. Acta,* 988, 427, 1989.
57. **Huang, K. S., Li, S., and Low, M. G.,** Glycosylphosphatidylinositol-specific phospholipase D, *Methods Enzymol.,* 197, 567, 1991.
58. **Rhode, H., Hoffmann-Blume, E., Schilling, K., Gehrhardt, S., Gohert, A., Buttner, A., Bublitz, R. R., Amme, G. A., and Horn, A.,** Glycosylphosphatidylinositol-alkaline phosphatase from intestine as a substrate for glycosylphosphatidylinositol-specific phospholipases-microassay using hydrophobic chromatography in pipet tips, *Anal. Biochem.,* 231, 99, 1995.
59. **Cockcroft, S.,** Phospholipase D: regulation by GTPases and protein kinase C and physiological relevance, *Prog. Lipid Res.,* 35, 345, 1997.

Chapter **2**

Analyzing the Sphingomyelin Cycle: Protocols for Measuring Sphingomyelinase, Sphingomyelin, and Ceramide

Heather Hayter and Yusuf A. Hannun

Contents

0-8493-3383-0/99/$0.00+$.50
© 1999 by CRC Press LLC

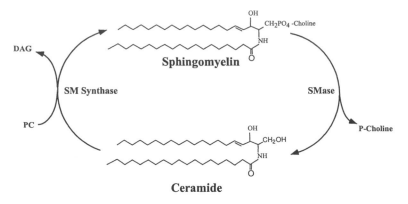

FIGURE 1

Sphingomyelin cycle. Upon addition of stimulus, SMase is activated, resulting in the production of choline–phosphate and ceramide. The cycle is completed by reincorporating ceramide into SM via SM synthase.[1]

I.   Introduction

Sphingolipids play important roles in apoptosis, differentiation, senescence, proliferation, and inflammation.[1] The sphingomyelin (SM) cycle (Figure 1), with the key components of SM, sphingomyelinase (SMase) and ceramide, has received substantial attention to date because of its role in cell signaling. Given the broad scope of the fields in which the SM cycle has been implicated, this chapter provides protocols that are intended to give the nonlipid biochemist a direct entry into studying this intriguing field. Each method provides information on not only the steps, but also the potential pitfalls both in the protocol and in interpretation. To understand potential problems, it is also important to understand the pathways of sphingolipid metabolism, which will help to select the appropriate assays for particular questions. Several excellent reviews of the field exist, and it is highly recommended that the reader consult one or more of them.[1-4]

The first step in sphingolipid synthesis is the condensation of palmitoyl-CoA and serine (Figure 2). Eventually dihydroceramide is converted to ceramide, and then SM synthase transfers a choline–phosphate from phosphatidylcholine to ceramide to produce SM. Ceramide is also the starting point for the synthesis of several other sphingolipids, including sphingosine, ceramide–phosphate, and glycosphingolipids.[5] Additionally, ceramide can be formed by the breakdown of other sphingolipids including SM and by the reacylation of sphingosine.[4] These metabolic interconversions complicate any straightforward analysis of data (see Figure 2).

Despite this complexity, the activity of the SM cycle can be monitored by measuring the components involved (SM, SMases, and ceramide.) Such data will implicate not only a sphingolipid source for ceramide, but would also suggest specific enzymes involved in regulating SM hydrolysis or ceramide generation.

The following protocols for measuring SMase, SM, and ceramide can be performed with very little special equipment. Additional background on general lipid techniques is readily available in several excellent volumes.[6-8]

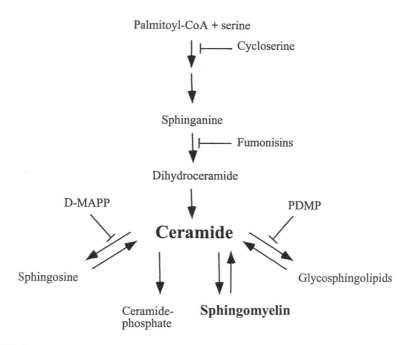

FIGURE 2

Sphingolipid synthesis and metabolism. An overview of sphingolipid synthesis and metabolism is presented. Where appropriate, inhibitors for the various steps are marked. See Table 1 for more information on the inhibitors. Note that several steps in the pathway are reversible.[1,4,5]

For additional information regarding the molecular species of sphingolipids, most investigators utilize high performance liquid chromatography (HPLC). These protocols will not be discussed in this chapter, but included in the references are several methods for measuring key sphingolipid species by HPLC for those laboratories with this capability.[9,10]

The methods to be described are most effective when applied to tissues and cultured cells. Plant and yeast samples can be analyzed, but require modification of the extraction technique.[6]

For a brief overview of useful inhibitors that provide information about lipids other than SM and ceramide, see Table 1. In addition, other protocols can be found at the Web site of the authors' laboratory, www.musc.edu/bcmb/ceramide/.

II. Protocol

A. Analyzing SMase

There are three primary SMases: neutral Mg^{2+}-dependent, neutral Mg^{2+}-independent, and acidic.[11] Other types of SMase have been reported, including zinc dependent and alkaline, but this chapter will concentrate on an assay capable of measuring the first three, as they are the ones typically implicated in signaling. Acidic SMase is

TABLE 1
Inhibitors of Sphingolipid Synthesis

Sphingolipid	Inhibitors
Glycosphingolipids	threo-1-phenyl-2-decanoylamino-3-morpholino-1-propanol (PDMP) blocks synthesis; ceramide and sphingosine levels increase
Ceramide	Fuminisin B_1 blocks *de novo* ceramide formation sphinganine accumulates
	D-erythro-N-myristoyl-aminophenol-1-propanol (D-MAPP) inhibits ceramidase; ceramide accumulates
All sphingolipids	Cycloserine blocks the initial condensation reaction

localized primarily in the lysosome, whereas the neutral Mg^{2+}-independent SMase is primarily cytosolic, and the Mg^{2+}-dependent SMase is found in the membrane.[11] The activity of each SMase varies greatly, depending on cell type, although acidic SMase is usually present in three to five times greater activity than the neutral SMases. Typically, the Mg^{2+}-independent SMase exhibit the lowest activity. Of these, only the acidic SMase has been cloned and purified.[12]

1.   SMase assay[13]

1.   Cell harvest and homogenization

   a.   Harvest cells. 3 to 5 million cultured cells or 1 g tissue required.

   b.   Resuspend in appropriate lysis buffer (acid or neutral). Homogenize by three freeze/thaw cycles or Dounce homogenization.

   c.   Spin homogenate at $3000 \times g$ for 10 min to remove nuclear debris.

Note: *As an option, follow by spinning supernatant at 100,000 × g for cytosol and membrane fractions.*

2.   Substrate preparation

   a.   Prepare substrate [^{14}C]-SM (can be obtained commercially from several sources including Amersham): 10 nmol [^{14}C]-SM, 1×10^5 dpm/sample. Dry down mixture under a stream of N_2 and resuspend in assay buffer. Set aside an aliquot for total counts and determination of specific activity (cpm/nmol).

3.   Protein assay

   a.   Perform protein assay on pellet and supernatant fractions using an equal volume of assay buffer as an assay blank. Subtract the blank value from cell/tissue sample values.

4.   Reaction preparation and initiation

   a.   Aliquot 100 µg of protein per sample, and bring volume up to 50 µl. Include a reagent blank containing every constituent except cell/tissue protein.

   b.   Add 50 µl of substrate.

   c.   Incubate for 15 to 30 min at 37°C.

   d.   Stop reaction by adding 1.5 ml $CHCl_3$:methanol (2:1, v/v).

5. Folch extraction
    a. Add 200 µl of H$_2$O and vortex well.
    b. Spin at 2000 × g for 5 min.
    c. Count 200 µl of the upper phase in a liquid scintillation counter.
6. Data analysis
    a. For data analysis, convert raw counts per minute to nanomoles, and then normalize to milligrams of protein per minute.

Assay notes

The neutral enzymes are not very stable, so assays should be performed the day of collection. Acidic SMase is more stable, and the authors have been able to keep samples frozen in lysis buffer at −80°C for up to a week and still retain activity. Also, for initial assays it is imperative to monitor the products of the reaction (choline phosphate and ceramide). To do this, dry down both the aqueous and organic phases of the Folch extraction. Thin-layer chromatography (TLC) on silica gel G plates is performed with aqueous standards—choline and choline phosphate—and organic standards—SM, ceramide, phosphatidylcholine (PC), and lyso-SM. Develop the TLC plate for aqueous samples using methanol: 0.5% NaCl:NH$_4$OH (100:100:2); organic samples using CHCl$_3$:MeOH:acetic acid:H$_2$O (5:3:0.8:0.5).

Considerations

Interpretation of these results should be done carefully, since changes in SMase activity can be a reflection of several possibilities. Thus, an increase in SMase activity may represent (1) transcriptional increase in the amount of enzyme, (2) a change in allosteric regulation, or (3) a post-translational modification. However, this alteration in SMase activity is neither necessary nor sufficient to document the role of a particular SMase. For example, a certain SMase may show a change in activity in enzyme assays without being responsible for the observed hydrolysis of SM in cells. Conversely, no apparent change in SMase activity can mean two things. First, SMase may not be involved in this particular pathway. Second, SMase may be involved, but its activity is regulated *in vivo* by an allosteric regulator that is lost during homogenization. Deducing the precise role of an SMase in a given pathway requires access to inhibitors that are not currently available.

B.   Sphingomyelin Measurement

Because so few of the SMases have been cloned and purified and because a change in SMase activity does not establish activation of the enzyme in cells, SM hydrolysis is the gold standard for determining SMase activity in cells. Ceramide measurements are complementary because ceramide can be derived from multiple sources and is metabolized in multiple ways. Therefore, any SMase activity assay should be supported by SM measurements.

1. SM measurement

The classic and most common method of measuring SM requires isotopic labeling of the choline head group with [³H]choline, extracting the lipids, and quantifying SM with a TLC method or a protocol using bacterial SMase.[14,15] An SM mass measurement protocol is also available.[15] Other assays include some that label with fluorescent SM or ceramide or use [³H]palmitic acid rather than choline. For radioactive labeling purposes, removing free choline is much easier to achieve than removing the free palmitic acid. Labeling with palmitic acid also labels all sphingolipids and most glycerolipids, making TLC isolation cumbersome. Choline is more specific because it is found only in sphingomyelin, PC, and lyso-PC.

The major difference between the two labeling methods discussed here (TLC and bacterial SMase) is time to completion. For the initial characterization of a cell line, it is recommended that the TLC assay be used to confirm results. Once a basic knowledge of the system is obtained, the bacterial SMase method is much more rapid, and can be readily used. The mass method is recommended in instances where it is desirable to know the absolute quantity of SM. For example, it is important to ascertain whether or not the amount of ceramide generated in a system can be accounted for by mass of SM hydrolyzed, and in this case one needs to determine SM mass. For both labeling methods, most variability derives from differential labeling of the cells, rather than from the assay itself.

2. SM hydrolysis, TLC method[16]

1×10^6 cells required.

1. Cell harvest
   a. Label cells with 1.0 to 0.5 μCi [³H]choline for 48 to 72 h.
   b. Harvest cells as described below.
2. General protocol for harvesting lipids from cultured cells
   Suspension cells or tissue:
   a. Wash cells once in phosphate-buffered saline (PBS).
   b. Add 3 ml of $CHCl_3$:MeOH (1:2) and proceed with Bligh–Dyer extraction (see below)
   Adherent cells—Option 1:
   a. Scrape cells with rubber policeman in 800 μl H_2O.
   b. Add 3 ml of $CHCl_3$:MeOH (1:2) and proceed with Bligh–Dyer extraction (see below) beginning at step c.
      OR
      Lyse cells by ultrasonication and remove 100 μl for protein determination. Then proceed with Bligh–Dyer (remember to add additional water if sonicating in a small volume) beginning at step c.
   Adherent cells—Option 2:
   a. Scrape cells with 1 ml MeOH.
   b. Wash plate with additional 1 ml MeOH; add to previous wash.

      c. Add 1 ml CHCl$_3$ and vortex until a monophase is achieved (add additional methanol if necessary).

      d. Proceed with Bligh–Dyer, beginning at step b.

3.   Bligh–Dyer extraction of total lipids[23]

      a. Add 3 ml CHCl$_3$:MeOH (1:2) to cell pellet.

      b. Add 800 µl H$_2$O; vortex well. Should be a monophase; if not, add methanol.

      c. Spin 2000 × g for 5 min to remove large debris.

      d. Transfer to fresh tube.

      e. Add 1 ml CHCl$_3$ and vortex.

      f. Add 1 ml H$_2$O and vortex.

      g. Spin 2000 × g for 5 min.

      h. Aspirate most of top, aqueous phase.

      i. Collect lower, organic phase.

      j. Dry down under nitrogen, or use a Speed Vac evaporator.

Note: *This is a good stopping point. Lipids may be stored at –20°C for up to 1 month.*

4.   Base hydrolysis and reextraction

      a. Add 250 µl methanolic NaOH (2 N) to the dried lipid fraction (methanolic acid and base are prepared by making a solution in MeOH rather than H$_2$O).

      b. Incubate 2 h at 37°C.

      c. Neutralize base with 250 µl HCl (2 N).

      d. Reextract lipids as for Bligh–Dyer extraction method, but substitute 430 µl H$_2$O, 500 µl of CHCl$_3$:MeOH (2:1), and 850 µl CHCl$_3$.

5.   TLC

      a. Resuspend dried lipid in 75 µl CHCl$_3$. Aliquot 10 µl for phosphate measurement, and spot 50 µl on TLC plate. Also, spot SM standards of 1 to 100 nmol SM.

      b. Develop plate in the following system: CHCl$_3$:MeOH:acetic acid:H$_2$O (50:30:8:5).

      c. Visualize SM using iodine vapors in a well-ventilated hood.

6.   Quantitation

      a. Quantitate SM spot via phosphoimager, or scrape the spot and quantitate by scintillation counting.

3.   SM hydrolysis, bacterial SMase method[14]

1.   1 × 10^5 cells required.

2.   Cell harvest and lipid extraction: duplicate steps 1 and 2 of TLC method.

3.   Preparation of mixed micelles

      a. Resuspend dried lipid in 50 µl of 0.19 M Tris-HCl, pH 7.4, 12 mM MgCl$_2$, and 0.2% Triton X-100.

   b. Vortex tubes vigorously. Sonicate with three 1-min bursts with a 1-min rest between bursts.

   c. Preincubate at 37°C for 5 min.

4. Incubation

   a. Add 50 μl of 2 units/ml SMase from *Streptomyces* sp. in 10 m*M* Tris-HCl, pH 7.4.

   b. Incubate for 30 min.

5. Extraction and quantitation

   a. Perform Folch extraction (see Section II.A.1, step 5).

   b. Count 400 μl of upper aqueous phase which should contain the labeled choline phosphate released from the SM by the action of bacterial SMase.

4.   SM mass measurement[18]

1×10^7 cells required.

For SM mass, no labeling is required. This eliminates the problem of differential labeling encountered in the previous methods. However, one must begin with a much larger cell number. SM levels can vary widely. A representative sample from HL-60s cells contains 50 pmol SM/nmol phosphate. As a percentage of lipid phosphate, SM can range from 4% in the liver to 25% in the pancreas.[17]

1. Cell harvest and lipid extraction

   a. Harvest cells.

   b. Extract lipid via Bligh–Dyer.

   c. Resuspend lipids in 1 ml $CHCl_3$. Aliquot 800 μl for SM measurement and 100 μl for phosphate measurements (see ceramide section for protocol).

2. Base hydrolysis

   a. Bring total volume for SM measurement to 1 ml. Add 100 μl of 2 *N* NaOH.

   b. Incubate at 37°C for 1 h.

   c. Neutralize with 100 μl of 2 *N* HCl.

   d. Perform Bligh–Dyer extraction (see Section II.B.2, step 3); begin by forming a monophase with 2 ml methanol and 600 μl of H_2O.

3. TLC

   a. Resuspend dried lipid in 75 μl $CHCl_3$, aliquot 10 μl for phosphate measurement, and spot 50 μl on a TLC plate.

   b. Spot SM standards of 1 nmol.

   c. Develop plate in $CHCl_3$:MeOH:acetic acid:H_2O (50:30:8:5).

   d. Visualize lipids with iodine vapor.

4. Scraping and extracting SM

   a. Scrape SM spots.

   b. Extract SM from silica by washing with $CHCl_3$:methanol (2:1) two times. Wash once with $CHCl_3$:methanol (1:2).

   c. Quantitate SM with phosphate assay.

Assay notes

For all three assays, 10% or more hydrolysis is considered significant since the mass of SM in cells usually far exceeds the mass of ceramide. Thus, a small quantity of hydrolyzed SM may generate a significant change in ceramide.

Considerations

A significant decrease in SM levels can imply several things. The most likely explanation, and one documented in mammalian cells, is activation of SMase resulting in SM hydrolysis. Another SMase, SMase D, which has not been observed in mammalian cells, could also cleave SM producing ceramide phosphate and choline. Alternatively, the decrease in SM may reflect a downregulation in SM synthase activity, which would concomitantly increase ceramide and decrease diacylglycerol (DAG).

C. Ceramide Measurement

Ceramide has been shown to function as a second messenger or lipid mediator in multiple systems, and at least two specific targets, a phosphatase and kinase, have been identified.[1] This lipid can be produced in several ways. The primary source of stimulus-induced ceramide formation appears to be sphingomyelinase, but *de novo* synthesis has been described.[1] A major complicating factor is that further metabolism of ceramide can occur quickly, either via reincorporation into SM or by hydrolysis into the free long-chain base. Thus, changes in ceramide levels provide evidence of sphingolipid involvement in a pathway, but do not implicate a specific enzyme without further analysis of SM levels or SMase activity. In addition, some inhibitors are available to determine which biochemical pathway is important (see Table 1 and Figure 2).[5]

1. Ceramide assay

As with SM, there are several methods available for ceramide quantitation. The authors prefer using the diacylglycerol kinase (DGK) method since this assay is very sensitive and quantitates the total mass of ceramide.[19,20] However, some basic tenets of this assay must be followed (see assay notes). With this method, total lipid is exposed to DGK in a mixed micelle system, resulting in the phosphorylation of DAG and ceramide, which can be separated by TLC. Most other methods involve prelabeling cells with palmitic acid or fluorescent analogs. While these methods can yield important results, differential labeling can produce large variability, making replication difficult. The DGK method avoids that variability and typically produces more consistent results. It also allows analysis of DAG, which is intimately connected to the SM cycle (see Figure 1).

1.   Ceramide measurement[19,20]
     1×10^6 cells required

2. Harvest cells and extract lipids
    a. Harvest cells.
    b. Extract total lipid via Bligh–Dyer method (see Section II.B.2, step 3) and dry lipid.
    c. Resuspend lipid in 75 μl CHCl$_3$. Aliquot 25 μl for phosphate determination and 25 μl for assay.
3. Prepare micelles, standards, and samples
    a. Preparation of BOG/DOPG micelles
        1. Add 5 g of octyl-β-D-glucopyranoside (BOG) to 20 ml acetone.
        2. Heat to 40°C in a water bath and filter using a glass fiber filter and vacuum filtration.
        3. Slowly add 100 ml diethyl ether to the solution and reheat if necessary to dissolve.
        4. Keep at –20°C overnight to allow crystals to form.
        5. Filter crystals. Wash with ice-cold diethyl ether.
        6. Weigh and store at –20°C.
        7. Aliquot appropriate amount of 1,2-dioleoylglycero-3-phospho-1-glycerol (DOPG) (Avanti Lipids #840475) into glass tubes (0.97 ml of 20 mg/l per tube).
        8. Dry under a stream of N$_2$; a thin film will appear on the bottom of the tube.
        9. Add 1 ml of a 7.5% solution of the recrystallized BOG in dH$_2$O.
        10. Store overnight at 4°C. The following day, vortex and sonicate until the lipid is dissolved.
        11. Store at –20°C for up to 1 year.
    b. Dry down ceramide and DAG standards, 10 to 160 pmol.
    c. Add 20 μl of BOG/DOPG mixed micelles. Vortex vigorously.
    d. Sonicate if necessary to resuspend lipid completely. This is a crucial step.
4. DGK reaction and lipid reextraction
    a. DGK reaction mixture (per sample). Prepare fresh on day of assay:

2X buffer:	50 μl
DTT:	0.2 μl
DGK solution:	1 μg/tube or 2.5 μg/tube (an excess of enzyme is required to provide complete phosphorylation of ceramide; see assay notes)
Dilution buffer:	Dilute reagents to a final volume of 70 μl.

    b. Add 70 μl of the DGK reaction mixture to each sample.
    c. Initiate reaction with 10 μl of 10 mM [γ-^{32}P]ATP (4.5 μCi).
    d. Incubate 30 min.
    e. Stop reaction with 3 ml of CHCl$_3$:MeOH (1:2).
    f. Perform Bligh–Dyer (Section II.B.2, step 3) and dry down organic phase (lipid). Upper phase contains most of the water-soluble radioactivity.
5. TLC
    a. Spot one half to one third of lipid on silica gel G TLC plate.
    b. Develop plate in CHCl$_3$:acetone:MeOH:acetic acid:H$_2$O (10:4:3:2:1).

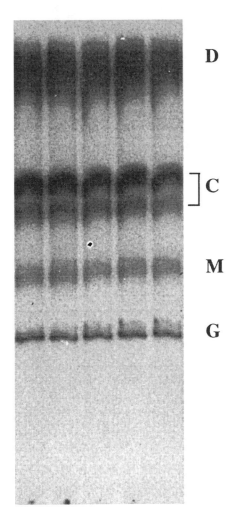

FIGURE 3

TLC plate after DGK analysis of lipid. This is a typical film taken from a DGK of total lipid from cultured cells. From the top, identification of bands follows: D, phosphatidic acid; C, ceramide–phosphate doublet; M, monoacylglycerol-phosphate; G, glycerol-phosphate; O, origin. The M and G bands do not appear in every DGK. Also, the intensity of each band of the ceramide doublet can vary.

    c. Visualize using phosphoimager or scrape spots and count by scintillation spectrometry.

    d. Normalize results to phosphate or protein content.

Once the plate is dried, the phosphorylated ceramide and DAG spots can be visualized on film after an overnight exposure. Ceramide-phosphate is a doublet running approximately in the center of the plate, with phosphorylated DAG (phosphatidic acid) running above it. Approximate R_f values are 0.52 and 0.54 for the ceramide doublet and 0.69 for DAG. Some other species occasionally observed include monoacylglycerol–phosphate (R_f 0.36), which runs below ceramide–phosphate. See Figure 3 for an example.

Quantitation of spots can be performed by scraping and counting or using a phosphoimager. Typically, control quantities of ceramide are in the range of 3 to 5 pmol ceramide/nmol phosphate. Treatments can increase ceramide levels over a wide range, with the greatest changes being approximately tenfold.

2.   Phosphate assay[21]

1.   Prepare standards and aliquot experimental lipid samples (unknowns).
2.   Add 100 μl 0.675 $Mg(NO_3)_2$ (ashing buffer) to each sample.
3.   Dry ethanol and chloroform from samples by heating at 80°C for 20 min.
4.   Flame samples using a Bunsen burner. Remove samples from flame when the brown gas ceases to be released. There should be a white precipitate at the base of the tubes. If flamed too long, the precipitate will appear brown or charred and the sample cannot be used.
5.   Add 300 μl of 0.5 N HCl to each sample and boil 15 min.
6.   To each tube add 600 μl ammonium molybdate and 100 μl 10% ascorbic acid. Vortex well.
7.   Incubate 30 min at 45°C or for 60 min at 37°C.
8.   Read at OD_{820}.

Assay notes

To achieve reliable and interpretable results, it is crucial to use a large excess of enzyme to drive the reaction to completion. For the recombinant pure preparation, 1 μg/tube is adequate. For membrane preparations of the enzyme, 2.5 μg/tube is recommended. In fact, it is imperative to perform a concentration-response analysis with the enzyme preparation and known amounts of lipid standards to ensure achievement of saturation and complete phosphorylation of ceramide and DAG.

Considerations

Three different sources can be responsible for an increase in ceramide levels. First, and most common, SMase can produce ceramide via SM hydrolysis. Second, ceramide can be formed as the result of *de novo* synthesis. Finally, an increase in ceramide can be derived from the reacylation of sphingosine. Determining the source of ceramide can be accomplished by measuring SM and by judicious use of inhibitors (see Table 1). Most increases in ceramide levels are measurable because of the relatively slow turnover rate of the SM cycle. A decrease in ceramide levels suggests catabolism or reincorporation; e.g., ceramide may be converted into sphingosine, ceramide–phosphate, SM, or the glycosphingolipids.

Reagents Needed _____

Neutral Lysis Buffer

> 25 mM Tris-HCl, pH 7.4
> 1 mM phenylmethylsulfonyl fluoride

5 mM dithiothreitol (DTT)
0.1 mM sodium molybdate (NaMoO$_4$)
750 μM ademosine triphosphate (ATP)
2 mM ethylenediaminetetraacetic acid (EDTA)
10 mM MgCl$_2$
0.1 mM sodium vanadate (Na$_3$VO$_4$)
10 mM β-glycerophosphate
0.2% Triton X-100
20 μg/ml each of chymostatin, leupeptin, antipain, and pepstatin

Acid Lysis Buffer

25 mM Tris-HCl, pH 7.4
1 mM phenylmethylsulfonyl fluoride
20 μg/ml each of chymostatin, leupeptin, antipain, and pepstatin
0.2% Triton X-100

Three Sphingomyelinase Assay Buffers

1. Neutral, Mg^{2+}-independent

20 mM Tris-HCl, pH 7.4
10 mM EDTA
0.05% Triton X-100

2. Neutral, Mg^{2+}-dependent

20 mM Tris-HCL, pH 7.4
5 mM MgCl$_2$
0.05% Triton X-100

3. Acidic

100 mM sodium acetate, pH 5.0
0.05% Triton X-100

2X Buffer

10 ml 0.5 M imidazole (pH 6.6)
0.21 g LiCl
1.25 ml 1 M MgCl$_2$
1.0 ml 0.1 M ethylemeglycol-bis-aminoethylether-tetraacetic acid (EGTA) (pH 6.6)
Dilute to 50 ml with distilled H$_2$O

Dilution Buffer

1 ml 0.5 M imidazole (pH 6.6)

2.5 ml 20 mM diethylenetriaminepentaacetic acid (DTPA)
Dilute to 50 ml with distilled H_2O

Dithiothreitol (DTT), 1 M

DGK Enzyme Solution

Calbiochem, recombinant pure No. 266724
Or membrane preparation No. 266726
Or Biomol membrane preparation No. SE-100

Ashing Buffer

0.675 M $Mg(NO_3)_2$ in ethanol
0.5 N HCl
1 N sulfuric acid

Ammonium Molybdate

0.42% ammonium molybdate in 1 N H_2SO_4

Ascorbic Acid

10% ascorbic acid in dH_2O, w/v, made fresh for assay

TLC Plates

Silica gel G plates (Whatman, LK60 silica gel No. 4865-821)

III. Discussion

Sphingolipid research is a rapidly expanding field. The complex catabolism and metabolism that the cell exploits for highly regulated signaling molecules enriches this field of novel regulated pathways, novel second messengers, and bioactive lipids. Activation of SMases and hydrolysis of SM initiate important regulatory pathways through the formation of ceramide and possibly other metabolites. Ceramide, in turn, regulates cell senescence, apoptosis, cell cycle, and inflammatory responses.[1,2,4] In this emerging field, it is important to consider methodology and interpretation of results carefully. This chapter attempts to clarify interpretation by discussing the implications of different results, and how hypotheses can be confirmed or extended using various methods.

At a basic level, it is critical to be certain that the methods utilized provide reliable data. This is most straightforward with SM measurements. For ceramide quantitation, the DGK assay is highly reliable, sensitive, and reproducible if used appropriately. The authors believe the most common source of error is not achieving total phosphorylation of ceramide, rendering results subject to various artifacts of enzyme (DGK) activity. With SMases, the assays indicate changes in enzyme amount and/or activity.

At another level, one must interpret the significance of the results. A fall in SM levels most likely implicates the activation of an SMase rather than inhibition of SM synthases. If the fall in the SM levels is accompanied by an increase in ceramide levels, then the most likely interpretation is that an SMase has been activated. The specific SMase responsible for this effect can be deduced from these criteria:

1. Is there a change in the activity of a particular SMase? This shows a link between agonist and a particular SMase, but, as discussed in Section II, it is neither necessary nor sufficient to implicate or rule out a role for a particular SMase.

2. Do inhibitors of a particular SMase inhibit the fall in SM or increase in ceramide levels? Unfortunately, there are no good pharmacological inhibitors of these SMases. Cells from Nieman–Pick patients or mice deficient in acid SMase can be used for evaluating the role of acid SMase.

3. What are the mechanisms involved in regulating SM hydrolysis? At this point, there is only one mechanism involved in the regulation of SMase, namely, the role of reduced glutathione (GSH) as an inhibitor of neutral SMase.[22] Therefore, a fall in GSH levels preceding SM hydrolysis suggests a role for neutral SMase.

It is hoped that the development of these additional methods and future availability of specific inhibitors will allow for more-sophisticated analysis of this pathway.

Acknowledgment

Supported in part by National Institutes of Health (NIH) Grant GM-43825.

References

1. **Hannun, Y. A.,** Functions of ceramide in coordinating cellular responses to stress, *Science*, 274, 1855, 1996.
2. **Ballou, L. R., Laulederkind, S. J., Rosloniec, E. F., and Raghow, R.,** Ceramide signalling and the immune response, *Biochim. Biophys. Acta*, 1301, 273, 1996.
3. **Kolesnick, R.,** Signal transduction through the sphingomyelin pathway, *Mol. Chem. Neuropathol.*, 21, 287, 1994.
4. **Jayadev, S. and Hannun, Y. A.,** Ceramide: role in growth inhibitory cascades, *J. Lipid Med. Cell Signal.*, 14, 295, 1996.
5. **Hannun, Y. A. and Jayadev, S.,** The sphingomyelin cycle: The flip side of the lipid signaling paradigm, in *Advances in Lipobiology*, Vol. 2, JAI Press, Greenwich, CT, 1997, 143.
6. **Hemming, F. W. and Hawthorne, J. N.,** *Lipid Analysis*, BIOS Scientific Publishers, Oxford, 1996.
7. **Vance, D. E. and Vance, J. E., Eds.,** *Biochemistry of Lipids, Lipoproteins, and Membranes*, Elsevier Science, Amsterdam, 1996.

8. Hamilton, R. J. and Hamilton, S., Eds., *Lipid Analysis: A Practical Approach*, Oxford University Press, New York, 1992.

9. Merrill, A. H., Wang, E., Mullins, R. E., Jamison, W. C., Nimkar, S., and Liotta, D. C., Quantitation of free sphingosine in liver by high-performance liquid chromatography, *Anal. Biochem.*, 171, 373, 1988.

10. Jungalwala, F. B., Hayssen, V., Pasquini, J. M., and McCluer, R., Separation of molecular species of sphingomyelin by reversed-phase high-performance liquid chromatography, *J. Lipid Res.*, 20, 579, 1979.

11. Liu, B. and Hannun, Y. A., Sphingomyelinases in cell regulation, *Semin. Cell Dev. Biol.*, 8, 311, 1997.

12. Lansmann, S., Ferlinz, K., Hurwitz, R., Bartelsen, O., Glombitza, G., and Sandhoff, K., Purification of acid sphingomyelinase from human placenta: characterization and N-terminal sequence, *FEBS Lett.*, 399, 227, 1996.

13. Wiegmann, K., Schutze, S., Machleidt, T., Witte, D., and Kronke, M., Functional dichotomy of neutral and acidic sphingomyelinases in tumor necrosis factor signaling, *Cell*, 78, 1005, 1994.

14. Jayadev, S., Linardic, C., and Hannun, Y., Identification of arachidonic acid as a mediator of sphingomyelin hydrolysis in response to tumor necrosis factor α, *J. Biol. Chem.*, 269, 5757, 1994.

15. Jayadev, S., Hayter, H. L., Andrieu, N., et al., Phospholipase A_2 is necessary for tumor necrosis factor α-induced ceramide generation in L929 cells, *J. Biol. Chem.*, 272, 17196, 1997.

16. Andrieu, N., Salvayre, R., and Levade, T., Evidence against involvement of the acid lysosomal sphingomyelinase in the tumour-necrosis-factor and interleukin-1 sphingomyelin cycle and cell proliferation in human fibroblasts, *Biochem. J.*, 303, 341, 1994.

17. Merrill, A. and Jones, D., An update of the enzymology and regulation of sphingomyelin metabolism, *Biochim. Biophys. Acta*, 1044, 1, 1990.

18. Okazaki, T., Bielawska, A., Bell, R. M., and Hannun, Y. A., Role of ceramide as a lipid mediator of $1\alpha,25$-dihyroxyvitamin D_3-induced HL-60 cell differentiation, *J. Biol. Chem.*, 265, 15823, 1990.

19. Schneider, E. and Kennedy, E., Phosphorylation of ceramide by diacylglycerol kinase preparations from *Escherichia coli*, *J. Biol. Chem.*, 248, 3739, 1973.

20. Preiss, J., Loomis, C. R., Bishop, W. R., Stein, R., Niedel, J. E., and Bell, R. N., Quantitative measurement of *sn*-1,2-diacylglycerols present in platelets, hepatocytes, and *ras*- and *sis*-transformed normal rat kidney cells, *J. Biol. Chem.*, 261, 8597, 1986.

21. Ames, B. and Dubin, D., The role of polyamines in the neutralization of bacteriophage deoxyribonucleic acid, *J. Biol. Chem.*, 235, 769, 1960.

22. Liu, B. and Hannun, Y. A., Inhibition of the neutral magnesium-dependent sphingomyelinase by glutathione, *J. Biol. Chem.*, 272, 16281, 1997.

23. Bligh, E. and Dyer, W., A rapid method of total lipid extraction and purification, *Can. J. Biochem. Physiol.*, 37, 911, 1959.

Chapter **3**

Expression of the Human Prostaglandin Endoperoxide H Synthases: Measurement of Cyclooxygenase Activity and Inhibition by Nonsteroidal Anti-Inflammatory Drugs

David DeWitt

Contents

0-8493-3383-0/99/$0.00+$.50

I. Introduction

Experimental and clinical data collected over the last 25 years indicate that the anti-inflammatory, antipyretic, and analgesic properties of aspirin, ibuprofen, and other nonsteroidal anti-inflammatory drugs (NSAIDs) result from their ability to block prostaglandin (PG) synthesis. The evidence that supports this hypothesis is straight-forward: each of the structurally diverse members of this large family of compounds share only one common property—they inhibit one or both of the two enzymes responsible for prostaglandin synthesis, prostaglandin endoperoxide H synthase-1 and -2 (PGHS-1 and PGHS-2; or cyclooxygenase-1 and -2, COX-1 and COX-2) (Figure 1).

PGHSs catalyze the conversion of arachidonate to PGH_2 (Figure 1). These enzymes have two activities: (1) a cyclooxygenase activity that incorporates two molecules of oxygen into arachidonate to form PGG_2 and (2) a peroxidase activity that reduces the 15-hydroperoxide group of PGG_2 to form PGH_2. NSAIDs prevent prostaglandin synthesis by competing with arachidonate for binding at the cyclooxygenase active site.[1] Thus, the anti-inflammatory properties of drugs can be predicted from their ability to inhibit the cyclooxygenase activities of the PGHSs.

Before 1991, only one PGHS had been described, the isozyme now called PGHS-1, or COX-1.[2] In that year, the laboratories of Simmons[3] and Herschman[4] working independently, discovered genes that encoded for a homologue of PGHS-1 that is now called PGHS-2, or COX-2. While the PGHS-1 and PGHS-2 isozymes have nearly identical cyclooxygenase turnover rates and affinities for arachidonate ($K_m \approx 5 \ \mu M$),[5-7] they are regulated quite differently, and have unique and separate biological functions. PGHS-1 is constitutively expressed and produces prostaglandins that regulate acute cellular responses to hormones and other stimuli. In contrast, the PGHS-2 is not expressed in most tissues, but can be induced by growth factors and inflammatory cytokines.

It is now believed that PGHS-2 produces prostaglandins that regulate inflammation-related processes and, therefore, that this isozyme must be the relevant site of action responsible for most of the therapeutically beneficial actions of NSAIDs. This belief is supported by the observations that PGHS-2 is induced (by cytokines) at sites of inflammation, and that drugs that selectively inhibit PGHS-2, but not PGHS-1, are potent anti-inflammatory, antipyretic, and analgesic agents. In addition, anti-inflammatory glucocorticoids and cytokines attenuate induction of PGHS-2, but have no effect on PGHS-1 expression.[1] In contrast, inhibition of PGHS-1 is thought to result in the major side effects of NSAID use, particularly ulcers and other gastrointestinal toxicities. These findings have led to a widespread interest in exam-

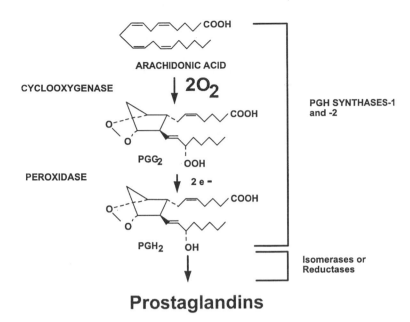

Prostaglandins

FIGURE 1

Reaction catalyzed by the prostaglandin endoperoxide H synthases-1 and -2. PGHS-1 and PGHS-2 catalyze two separate reactions: (1) a cyclooxygenase activity incorporates two molecules of oxygen into arachidonate to form PGG_2 and (2) a peroxidase activity reduces the 15-hydroperoxide group of PGG_2 to form PGH_2. NSAIDs prevent prostaglandin synthesis by competing with arachidonate for binding at the cyclooxygenase active site.

ining the selectivity of NSAIDs for inhibition of PGHS-1 and PGHS-2. Unfortunately, the human enzymes have no convenient tissue source, and are not commercially available, so *in vitro* expression systems have had to be employed for these studies.

Described below is an efficient and simple protocol for the expression of the human PGHS-1 and PGHS-2 enzymes in cos-1 cells, and three alternative protocols for examining inhibition of these recombinant enzymes by compounds of interest. The relative merits of each of these assay methods are discussed with respect to the ease of use, cost, and validity of results that can be obtained.

II. Protocol

A. Expression of PGHS-1 in Cos-1 Cells

1. Transfection of cos cells

Considerations. The protocols below employ human PGHSs enzymes. It is probably best to compare the human PGHS-1 and PGHS-2 enzymes, as species-to-species differences in NSAID potencies have been observed.[6,8] However, if the source of enzyme is unimportant, semipure ovine PGHS-1 and PGHS-2 are commercially available (Cayman Chemical and Oxford Biomedical). While recombinant human PGHS-2 can be purchased, it is expensive, and no complementary recombinant human PGHS-1 is yet available.

The number of inhibitor determinations to be performed, and whether or not tissue culture facilities are readily available, may also influence the decision to buy commercial PGHS or to use an *in vitro* expression system. For small numbers of assays (10 to 20), it is less expensive to buy PGHSs than to do the transfections. Conservatively, it costs about $2 for labor and materials to grow and transfect a 100-mm tissue culture plate, and isolate membranes. The cost is about 50% higher than that incurred by buying commercial ovine PGHS.

On average, about 100 to 150 units of cyclooxygenase activity are obtained per 100-mm tissue culture plate (2×10^6 cells) (1 unit = amount of enzyme that converts 1 nmol of O_2/min at 37°C). At the most sensitive setting (100-mV full-scale setting on the chart recorder), about 50 to 100 units of activity are required for every oxygen electrode activity determination. While the oxygen electrode can accurately measure as little as 10 units of activity, it is useful to start with about 100 units so that an accurate inhibitor dose–response curve can be constructed. Similar numbers of transfected cells are required when measuring product formation using either the oxygen electrode or using ^{14}C-arachidonate, while only about 1/5-1/10 as much enzyme is required for the enzyme-linked immunsorbent assay (ELISA)/radioimmunoassay (RIA) protocol. Thus, 10-15 plates (1000-1500 units) of transfected cos-1 are required to determine one IC_{50} for one PGHS isozyme using the oxygen electrode or ^{14}C-AA assays, while 1-2 plates (100 to 150 units) are sufficient for determining a dose-response curve for one enzyme using the ELISA/RIA protocol.

Procedures

1. Grow cos-1 cells (ATCC CRL-1650) in Dulbecco's modified Eagles medium (Gibco-BRL 11965-092) containing 10% fetal calf serum, 10 U/ml penicillin, and 10 µg/ml streptomycin (Gibco-BRL 15140-122) (DME-FCS) in a humidified incubator with 5% CO_2 at 37°C. Three days before transfection, subculture cells to new plates.

   a. Add 1 ml trypsin-EDTA (Gibco-BRL 25200-056) to each confluent plate of cos-1 cells.

   b. After 2 to 3 min, wash cells off the plate into DME-FCS.

   c. Centrifuge cells at $1000 \times g$ for 10 min, and replate at a concentration of 2×10^5 cells/100-mm tissue culture plate. At the time of transfection (day 3), cells should be just reaching 90 to 100% confluency (about 2×10^6 cells/plate).

2. Prepare plasmid DNA (pSVLN-PGHS-1 or pSVLN-PGHS-2) by cesium chloride (CSCI) centrifugation or anion exchange chromatography (Qiagen) and resuspend into TE buffer at a concentration of approximately 0.5 to 1.0 mg/ml. Sterilize DNA by filtering through a 0.22 µm Millex-GS syringe filter (Millipore; SLGS0250S) into a sterile microfuge tube.

3. Prepare 3 ml DEAE dextran–DNA solution for each 100-mm plate of cells to be transfected by adding 250 µg/ml DEAE dextran (Pharmacia Biotech;17-0350-01) to DME media without serum; sterilize by filtration (see above).

a. Add sterile plasmid DNA (5 μg/ml) to the DEAE dextran solution.

b. At this time, also prepare 7 ml DME-FCS containing 52 μg/ml chloroquin for each plate of cells to be transfected. Add chloroquin to the DME-FCS and sterilize by filtration with a 0.2 μm vacuum filtration unit (Nalgene; 121-0020).

4. Remove media from cos-1 cells. Tilt plates and allow media to pool for 1 to 2 min; then remove the final drops of media. Alternatively, wash the plates once with 5 ml phosphate-buffered saline (PBS) (Gibco-BRL; 20012-027).

5. Add 3 ml of DEAE dextran–DNA solution carefully to the edge of each tissue culture plate, and return them to the incubator for 1 h. Cos-1 cells are easily dislodged from the tissue culture plate, so all additions should be made gently.

6. Next, add 7 ml DME-FCS chloroquin solution to the DEAE dextran–DNA already on the plates, and return cells to the incubator for an additional 5 h.

7. Finally, remove the DNA- and chloroquin-containing media from the cells. Either tilt plates and drain to remove the final drops of media, or wash with 5 ml PBS.

8. Add 10 ml DME-FCS to each plate and return the plates to the incubator.

9. Harvest the transfected cells no earlier than 40 h later. The cells will appear granular and unhealthy from overexpression of PGHS, but should still be attached to the plate. If the oxygen electrode or the ^{14}C-arachidonic acid assays are to be conducted, remove media and scrape cells into 5 ml PBS using a large Teflon spatula, or a rubber policeman. For RIA or ELISA assays, trypsinize cells and collect as in step 2.

10. Centrifuge the cells at $1000 \times g$ for 5 min, discard the supernatant, and place cell pellets on ice.

2. Preparation of microsomal membranes

PGHSs are membrane proteins, and therefore preparation of microsomal membranes is a simple way to purify the enzymes partially before use in inhibitor assays. While not absolutely essential, isolation of membranes results in about a tenfold purification of PGHSs, and removes many of the soluble and nuclear proteins that may interfere with measurements of cyclooxygenase activity and determinations of inhibition constants.

Procedures

1. Resuspend cells in 5 ml 0.1 M TrisCl, pH 8.0, in a 12 to 15 ml plastic centrifuge tube. Sonicate cells three to four times for 20 s on ice. Alternatively, use a Dounce homogenizer to homogenize the cells thoroughly. The most efficient recovery of microsomal membranes is achieved by using at least five 100 mm-plates (1×10^7 cells).

2. Centrifuge the cell homogenate at $10,000 \times g$ for 10 min.

3. Transfer the supernatant to a clean 5-ml ultracentrifuge tube and centrifuge for 45 min in a Beckman SW-50.1 rotor or equivalent at 40K rpm ($100,000 \times g$).

4. Remove the supernatant and resuspend the microsomal pellet using a small Dounce homogenizer into 0.05 ml 0.1 M TrisCl, pH 8.0, per plate of cells. The microsomal pellet, which may be difficult to observe, will appear clear and gelatinous with a slightly brown tint. It is preferable to resuspend this pellet first, by pipetting up and down rapidly with a Pasteur pipette or a Pipetteman-type pipetter, and then to transfer the pellet to

an homogenizer. An alternative method, which is not as efficient, to resuspend the microsomal pellet is to draw the pellet up into a syringe using a small-bore needle. A thoroughly homogenized microsomal pellet makes activity measurements more reproducible. Reserve some of the homogenate for protein quantitation. Relative expression of the PGHS-1 and PGHS-2 protein can be measured using Western blot analysis of the cos-1 cell homogenates with commercially available isozyme-specific antisera (Cayman Chemical or Oxford Biomedical). Alternatively, the ^{14}C-arachidonate assay protocol (Section II.B) is a simple means to confirm that active PGHS has been expressed.

B.   Measurement of PGHS Activity and NSAID Inhibition

1.   Oxygen electrode determination of cyclooxygenase activity and inhibition

Measurements of cyclooxygenase activities are complicated by the fact that the PGHSs undergo rapid suicide inactivation. On average, each enzyme turns over no more than 1400 molecules of arachidonate before becoming inactivated. Thus, at saturating substrate concentrations, cyclooxygenase activity is linear only for only about 30 s (Figure 2). Because of this rapid inactivation, the most accurate determination of cyclooxygenase activity—and that most useful for determining enzymatic kinetic constants (K_m and K_i)—is the measurement of oxygen utilization during catalysis, which can be monitored using an oxygen electrode. A chart trace recorded during an activity determination using an oxygen electrode is presented in Figure 2. When arachidonate is added to the reaction mixture (Figure 2), there is a short characteristic lag as the PGHS-catalyzed production of lipid-peroxides builds up to a level necessary for full activation of the enzyme.[9] Once activated, PGHS rapidly incorporates O_2 into arachidonate to form PGG_2. The simultaneous decrease in O_2 in the reaction mixture is recorded as a sharp linear drop in the trace. The slope of this line is directly proportional to the activity of the enzyme. This line gradually begins to curve and flatten out as inactivation of the PGHS occurs. If the measurement shown in Figure 2 were continued, the trace would flatten horizontally as the enzyme lost all activity.

While oxygen electrode measurements represent the most accurate determination of cyclooxygenase activity, they require an investment in specialized equipment that may not be justified if only a few IC_{50} determinations are to be made. In this situation, assays that measure PGE_2 production using ELISA or RIA (Section II.B.3), or formation of ^{14}C-prostaglandin products (Section II.B.4), may be more practical. It is important to remember, however, that end-point assays, such as the ELISA or RIA, usually do not measure activities within the linear range of the uninhibited PGHSs and may, therefore, underestimate the potency of the compounds being tested.

Procedures

1.   Allow the oxygen electrode–water bath assembly to warm up to 37°C for 10 to 20 min prior to beginning measurements. Follow the manufacturer's instructions for the proper preparation of the electrode.

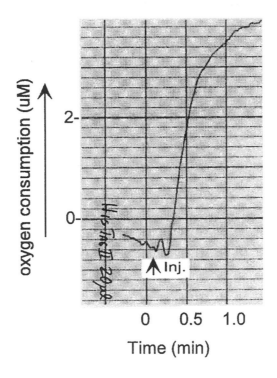

FIGURE 2
Oxygen electrode trace of the cyclooxygenase activity of human PGHS-2 expressed in cos-1 cells. Cyclooxygenase activity is measured by determining the slope of oxygen consumption within the first 30 s following injection (Inj.) of PGHS into the oxygen electrode cuvette. (One unit of activity is defined as the amount of enzyme that converts 1 nmol of O_2/min at 37°C.) In this example, the rate is 52 squares/min: full scale (100 spaces) at this recorder setting (10 mV) is 20 μM O_2. Thus, 10.4 μM O_2/min are consumed, which means that in the 3-ml reaction cuvette 31.2 nmol of oxygen are consumed, and there are 31.2 units of cyclooxygenase activity in the sample.

2. Add into each cuvette: a magnetic stirring bar, 3 ml Tris-phenol buffer, 5 μl 2 mg/ml arachidonate (final 10 μM), 20 μl 3.4 mg/ml hemoglobin, and up to 30 μl of vehicle or test drug. Insert the cuvettes into the bath assembly and secure. Turn on the circulating water bath, and begin stirring. Allow 5 min for the temperature to equilibrate.

3. Stop stirrer, insert electrode, remove all bubbles, then restart the stirrer. Adjust the recorder pen to the near full scale, start the recorder, and wait until a steady baseline is obtained. For most measurements with cos-1 microsomes it is necessary to make the most sensitive measurement possible. This can be accomplished by using the more sensitive ranges (100 to 200 mV) on the chart recorder. A voltage offset device (see Figure 5) is necessary when using these more sensitive scales, as this attachment allows one to adjust the chart recorder baseline before each measurement.

4. After the baseline is established, inject 25 to 100 μl microsomal membranes from the transfected cos-1 cells or 50 to 100 U of commercial enzyme, using a Hamilton syringe. To facilitate injection, attach a small piece of surgical tubing that fits snugly over the syringe needle and that is long enough to almost reach the stir bar. After a several second delay, there should occur a rapid deflection of the recording pen denoting a decrease in the dissolved O_2.

5. When the reaction begins to slow (1 to 2 min), turn off the recorder (or lift the pen), stop the cuvette stirrer, and remove the oxygen electrode from the first cuvette. Wash the electrode thoroughly with distilled water, return it to the second cuvette, and then restart the stirrer. When the baseline stabilizes, begin the next measurement.

6. Be sure to wash the cuvettes between uses with ethanol to remove arachidonate and some of the less-soluble NSAIDs that can bind to the glass and affect the accuracy of the measurements.

7. Note that, on average, about 100 to 150 units of cyclooxygenase activity are recovered per 100-mm tissue culture plate (2×10^6 cells) (1 unit = amount of enzyme that converts 1 nmol of O_2 per min at 37°C) with a specific activity of about 150 to 250 U/mg microsomal protein.

2. Instantaneous and time-dependent inhibition of cyclooxygenase activity

Inhibition by NSAIDs are most often reported as IC_{50} values (Figures 3A, B), the concentration of drug that is required to inhibit the activity of each enzyme by 50%. IC_{50} values are not inhibitor constants, but instead are variables that depend on substrate concentration and other assay conditions. Nevertheless, IC_{50} values require many fewer measurements to obtain than true inhibitor constants (K_i values), and thus are more commonly used. Since PGHS-1 and PGHS-2 have similar K_m values,[6-8] the IC_{50} values for the two isozymes can be compared directly to obtain an accurate estimate of the selectivity of a drug for PGHS-1 and PGHS-2.

Both instantaneous and time-dependent inhibition of cyclooxygenase activity can be measured using an oxygen electrode. The protocol described above—in which arachidonate and inhibitor are first placed in the oxygen electrode cuvette and then enzyme is added—is a measurement of instantaneous inhibition. Inhibition in the instantaneous assay is dependent only on competition between the inhibitor and arachidonate for the cyclooxygenase active site.

While measurement of instantaneous inhibition provides the most accurate estimate of the relative affinity of PGHS-1 and PGHS-2 for different drugs, such determinations are often not adequate predictors of the *in vivo* potency of a drug. This is because many NSAIDs, including all of the PGHS-2-selective drugs that have been identified, only achieve maximum potency when they have been incubated with PGHS for a period of time. (See below and Reference 10, for a description of inhibitor mechanisms.) At physiological concentrations, these inhibitors may not significantly inhibit PGHS activity when added simultaneously with arachidonate, but they often can substantially or completely inhibit activity when preincubated with PGHS for several minutes before addition of substrate. For these reasons, it is advisable to examine the time dependency of inhibition of any compound that has been found to have activity in the instantaneous inhibition assay. This can be accomplished by preincubating enzyme for 0.5 to 10 min with a concentration of inhibitor equal to about one tenth the IC_{50} value determined with the instantaneous assay. The enzyme–inhibitor mix can then be injected into the oxygen electrode cuvette. Activity will decline for time-dependent inhibitors with increasing incubation time. The dilution of inhibitor upon addition to the reaction cuvette is usually sufficient to eliminate any component of instantaneous inhibition.

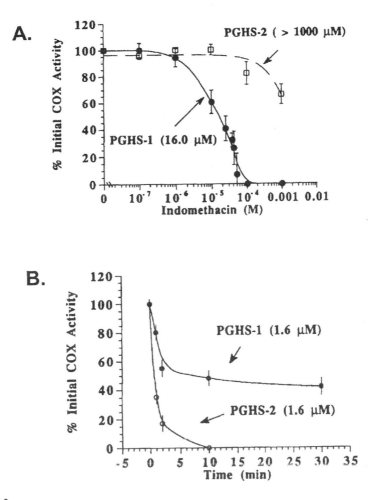

FIGURE 3

Characterization of the NSAID inhibition of the PGHS-1 and PGHS-2 isozymes-A. A typical dose–response curve measuring instantaneous inhibition of human PGHS-1 and PGHS-2 by indomethacin. B. Time-dependent inhibition of PGHS-1 and PGHS-2 by indomethacin. Note that indomethacin, at a concentration (1.6 μM) that does not significantly inhibit PGHS-1 and PGHS-2 in the instantaneous assay, can partially or completely inhibit these enzymes following preincubation.

3.   ELISA or RIA of cyclooxygenase activity

Considerations.   This protocol is adapted from a method first described by Cromlish and Kennedy[11–13] that employed baculovirus-infected spodotera cells expressing human PGHS-1 and PGHS-2. The relative level of expression of PGHS in insect cells is similar to that detected in cos-1 cells, so either cell system can be used. However, PGHS-1 is expressed more efficiently in cos-1 cells. The main advantage of using an ELISA or RIA to measure PGHS activity compared with using an oxygen electrode is that about one tenth to one fifth the amount of enzyme is needed for the former assays. Only 1 to 2×10^5 cells or 5 to 10 U of PGHS activity are needed for each activity measurement with an ELISA or RIA, compared

Lipid Second Messengers

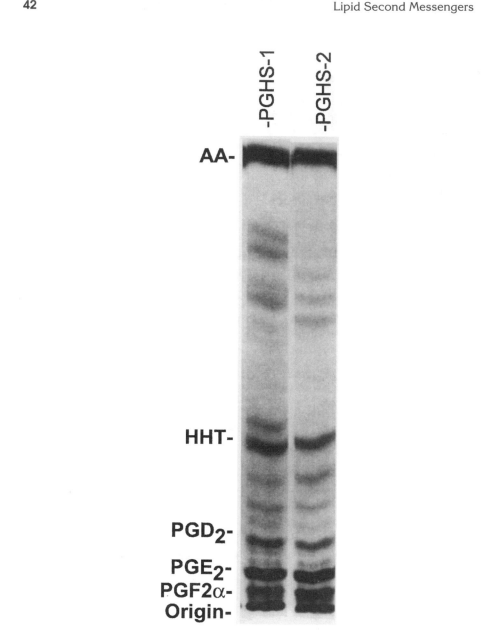

FIGURE 4

Typical autoradiograms of thin-layer chromatographic (TLC) separations of the products derived from incubation of human PGHS-1 and PGHS-2 with ^{14}C-arachidonate. Autoradiogram of TLC plate on which prostaglandin products produced in the ^{14}C-arachidonate assay have been separated (Section II.B.4). The relative production of prostaglandins can be determined by densitometric analysis of the autoradiogram or, alternatively, by scintillation counting of the regions of the silica gel corresponding to the products and arachidonate.

with 1 to 2×10^6 cells or 50 to 100 U of cyclooxygenase activity that are needed for an oxygen electrode measurement. In addition, the ELISA and RIA methods can be scaled up or automated if large numbers of compounds are to be evaluated.[11–13] However, the RIA requires a scintillation counter, and the ELISA assay requires a microtiter plate absorbance reader. In addition, the RIA and ELISA assay kits are costly. The protocol described below involves a preincubation step, so full inhibition by time-dependent inhibitors will be observed. It should be repeated, however, that it is impractical to run these assays within the linear time range for cyclooxygenase activity, so the IC_{50} values obtained may underestimate the potencies of compounds tested.

Procedures

1. Resuspend cells from each 100-mm tissue culture dish (2×10^6 cells) in 2.0 ml Hank's buffered salt solution (HBSS) (Gibco-BRL 12175-095) containing 10 mM HEPES, pH 7.4. Separate into 0.2-ml aliquots in 96-well plates or in microfuge tubes.

2. Add 2 μl inhibitor or vehicle, and incubate for 10 min at 37°C.

3. Add 22 μl 0.1 mM arachidonate in PBS (final 10 μM), and incubate for an additional 10 min at 37°C.

4. Terminate reactions by adding 20 μl 1 N HCL. Neutralize reaction by adding 20 μl 1 N NaOH. Centrifuge 96-well plates for 10 min at $300 \times g$, and microfuge tubes for 1 min at 12,000 rpm.

5. Remove the supernatant and determine PGE_2 levels by RIA (Amersham, RPA-530; or NEN, NEK-070) or ELISA (Cayman Chemical). Expect between 100 to 400 picograms PGE_2 per uninhibited assay (2×10^5 cells).

4. Measurement of cyclooxygenase activity with [^{14}C]arachidonate

Considerations. The main advantage of the ^{14}C-arachidonate assay is that it does not require an oxygen electrode. This protocol, which was first described by Meade et al.,[6] is also useful to characterize the mixture of products that are synthesized by PGHS, or to simply confirm that the cos cells express active PGHS.[6] In addition, because this assay uses whole cells, it is possible to examine the reversal of time-dependent inhibition by resuspending drug-treated cells in medium that does not contain inhibitor. The disadvantages of this method are that a densitometer or a scintillation counter is required to quantitate activity, that the activity measurements are at best only semiquantitative, and that ^{14}C-arachidonate is expensive. Thus, for routine screening of drug inhibition, this protocol is the least preferred.

Procedures

1. Resuspend PGHS-transfected cos-1 cells in 1.0 ml DME per 100-mm tissue culture dish and separate into 0.5-ml aliquots in silanized glass tubes.

2. Incubate cells with 5–50 μl NSAID or vehicle alone for 10 min at 37°C.

3. Add 2.5 μl of ^{14}C-arachidonate (final 10 μ*M*) (40 to 60 mCi/mmol NEN Life Sciences; NEC661), and continue the incubations for 10 min at 37°C.

4. Centrifuge the cells and transfer the supernatant to a new tube. Add 0.5 ml ice-cold acetone, vortex, and centrifuge in a tabletop centrifuge (1000 × *g*) for 10 min.

5. Transfer the supernatants to a clean tube and acidify by adding 250 μl 0.1 *M* HCl. Add 2.5 ml chloroform, mix, and centrifuge again to separate the layers.

6. Discard the aqueous phase (top) and dry the organic phase under N_2.

7. Resuspend the lipid products in 100 μl ethanol and apply to a silica gel 60 thin-layer chromatography (TLC) plate (VWR). Apply 100 ng each of authentic unlabeled arachidonate, PGE_2, PGE_2, and $PGF_{2\alpha}$ standards (Cayman Chemical). Develop the TLC plate twice in benzene:dioxane:acetic acid:formic acid (82:14:1:1; v/v/v/v).

8. To visualize PG standards, place a small amount of glass wool in a Pasteur pipette, add a small amount of iodine to the pipette, and insert more glass wool on top of the iodine. Using compressed air, or a pipette bulb, in a fume hood, blow iodine vapors over the lanes containing the standards, which will stain a faint yellow.

9. Note that ^{14}C-labeled products can also be visualized by 24 to 48 h exposure to XAR-5 X-ray film (Kodak) at –80°C (Figure 4). Relative conversion of arachidonate to PG products can be quantified by densitometry of the film. Alternatively, using the autoradiograph, mark the positions of the major PGs, PGD_2, PGE_2, $PGF_{2\alpha}$, and arachidonate, and then scrape these areas into scintillation vials using a razor blade, and quantitate by scintillation counting.

Reagents Needed

Tris-Phenol, pH 8.0 (0.1 M TrisCl, 1 mM phenol)

> 12.1 gm TrisCl
> 94 μl phenol
> Add approximately 900 ml ddH$_2$O, adjust to pH 8.0 with HCl
> Bring to 1 l

TE Buffer

> 0.01 *M* TrisCl, pH 7.5
> 0.001 *M* EDTA

Arachidonate (2 mg/ml)

Pipette 20 μl of 100 mg/ml arachidonate acid (in ethanol) (Cayman Chemical) into a glass test tube and remove the ethanol by evaporation with a stream of nitrogen. Add 1 ml of Tris-phenol, vortex vigorously for 60 s. Arachidonate is not fully soluble at this concentration and will remain cloudy when fully suspended.

Hemoglobin (3.4 mg)

Add 3.4 mg hemoglobin (Sigma) to 1 ml of Tris-phenol

Nonsteroidal Anti-Inflammatory Drugs

Resuspended NSAIDs at 1 to 10 mM in dimethyl sulfoxide (DMSO) or ethanol. All vehicles should be tested in control assays.

Sources of Enzymes _____

Plasmids—pSVLN-PGHS-1$_{hu}$ and pSVLN-PGHS-2$_{hu}$

Plasmids vectors for the expression of human PGHS-1 and PGHS-2 in cos-1 cells—request from the author: dewittd@pilot.msu.eud

Prostaglandin H Synthase 1 (Ovine)

> Cayman Chemical (Cat. No. 60100)
> Oxford Biomedical (Cat. No. PG01)

Prostaglandin II Synthase 2 (Ovine)

> Cayman Chemicals (Cat. No. 60120)
> Oxford Biomedical (Cat. No. NP04)

Prostaglandin H Synthase-2 (Human) Recombinant Enzyme

> Oxford Biomedical (Cat. No. RP02)

Equipment _____

Oxygen Electrode

> Yellow Springs Instrument Standard Bath System (Model 5300)

Single-Channel Flatbed Chart Recorder

> Lenseis, Inc. (Princeton Junction, NJ, Model No. 250E-1) or equivalent

Voltage Offset Device

> See diagram (Figure 5). The output from the oxygen electrode is from 0 to 1 V corresponding to 200 μM O_2. However, during typical assays involving 50 to 100 U of cyclooxygenase activity, changes in O_2 concentrations on the order of 5 to 20 μM occur. Thus, it is necessary to use the chart recorder on the more sensitive 100 to 200 mV sensitivity scales. The voltage offset device, which is simply a 1.5 Volt battery and a rheostat, facilitates positioning the recorder pen when making these more sensitive measurements.

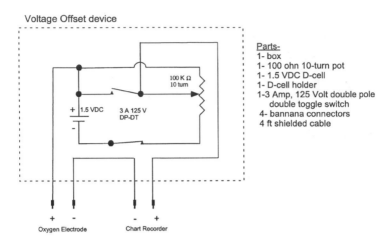

FIGURE 5

Circuit diagram for voltage offset device. This device in installed in parallel between the oxygen electrode output and the chart recorder. This device applies a constant voltage to offset the voltage output from the oxygen electrode so that the pen position can be adjusted to the top of the recording chart when more sensitive recording scales (100 to 200 mV) are employed. (The circuit is shown in the on position.)

Circulating Water Bath

Haake D8-L (or equivalent, to maintain operating temperature 37 ± 0.1°C)

Commercial Sources

Cayman Chemical Company
690 KMS Place
Ann Arbor, MI 48108 USA
Tel. 1-800-364-9897;
mrkting@caymanchem.com
http://www.caymanchem.com

Oxford Biomedical Research
P.O. Box 522
Oxford, MI 48371 USA
Tel.1-800-692-4633
Fax 248-852-4466
info@oxfordbiomed.com
http://www.oxfordbiomed.com

III. Discussion—Pharmacology of PGHS Inhibition

This chapter includes protocols for expressing PGHS-1 and PGHS-2 enzymes and assessing modes of inhibition by NSAIDs and other pharmacological agents. It has already been learned that NSAIDs inhibit by three pharmacologically distinct mechanisms, each of which can affect both the absolute potency and relative selectivity of a compound for PGHS-1 and PGHS-2, and thus the biological activity of these drugs. Aspirin, the oldest and most familiar NSAID, is unique in that it alone covalently modifies PGHS. Aspirin acetylates a serine residue in the active site of the PGHSs. *In vitro* mutagenesis experiments,[14,15] and X-ray crystallographic analysis of the aspirin-acetylated enzyme,[16] have established that when this active site serine is acetylated by aspirin the acetyl group protrudes into the cyclooxygenase active site and interferes with the binding of arachidonic acid. Covalent modifications of PGHSs by aspirin cause permanent inactivation of the enzymes. The ability of aspirin to modify PGHS-1 is the basis for the unique, long-lived effect of aspirin. This property also makes aspirin an effective antiplatelet agent because circulating platelets, unlike most cells, do not synthesize new PGHS-1. Inactivation of platelet PGHS-1 is the basis for reduced mortality from heart disease seen with regular aspirin use.[17–19]

Many common NSAIDs, such as ibuprofen and naproxin, are simply reversible competitive inhibitors of the cyclooxygenase.[8,20] However, there are a number of NSAIDs, including indomethacin, flurbiprofen, and meclofenamate, which exhibit an intermediate form of inhibitory behavior known as time-dependent inhibition.[8,20–22] Binding of these drugs to the PGHSs yields an initial enzyme–inhibitor (EI) complex typical of a reversible competitive inhibitor, but this EI complex slowly rearranges to a complex (EI*) from which the drug dissociates very slowly (minutes to hours).[10,21,22] With these compounds, inhibition only becomes maximal after several minutes incubation.

More recently, compounds have been identified that have mixed modes of inhibition. All of these have been found to be selective inhibitors of PGHS-2. These drugs are selective because they are time-dependent pseudo-irreversible inhibitors of PGHS-2 and simple competitive inhibitors of PGHS-1.[23–26] At physiological concentrations, these inhibitors do not significantly affect PGHS-1 activity but bind the cyclooxygenase active site of PGHS-2 tying up this enzyme in inactive EI* complexes. Included in this group of PGHS-2 selective agents are DuP697,[24] SC58125 and SC-58635 (celecoxib),[23,27] L-745-337,[28] NS398,[29] and etodolac.[30,31] These mixed-mode inhibitors are prototypes for drugs which are at present in clinical trial and which promise to be the first of a new generation of safer NSAIDs.

References

1. **Smith, W. L., Garavito, R. M., and DeWitt, D. L.,** Prostaglandin endoperoxide H synthases (cyclooxygenases)-1 and -2, *J. Biol. Chem.*, 271, 33157, 1996.
2. **DeWitt, D. L. and Smith, W. L.,** Primary structure of prostaglandin G/H synthase from sheep vesicular gland determined from the complementary DNA sequence, *Proc. Natl. Acad. Sci. U.S.A.*, 85, 1212, 1988.
3. **Xie, W., Chipman, J. G., Robertson, D. L., Erikson, R. L., and Simmons, D. L.,** Expression of a mitogen-responsive gene encoding prostaglandin synthase is regulated by mRNA splicing, *Proc. Natl. Acad. Sci. U.S.A.*, 88, 2692, 1991.

4. **Kujubu, D. A., Fletcher, B. S., Varnum, B. C., Lim, R. W., and Herschman, H. R.,** TIS10, a phorbol ester tumor promoter inducible mRNA from Swiss 3T3 cells, encodes a novel prostaglandin synthase/cyclooxygenase homologue, *J. Biol. Chem.*, 266, 12866, 1991.

5. **Laneuville, O. I., Breuer, D. K., Xu, N., Huang, Z. H., Gage, D. A., Watson, J. T., Lagarde, M., DeWitt, D. L., and Smith, W. L.,** Fatty acid substrate specificities of human prostaglandin H synthases-1 and -2, formation of 12 hydroxy-(9Z,13E/Z,15Z)-octadecatrienoic acids from alpha-linolenic acid, *J. Biol. Chem.*, 270, 19330, 1995.

6. **Meade, E. A., Smith, W. L., and DeWitt, D. L.,** Differential inhibition of prostaglandin endoperoxide synthase (cyclooxygenase) isozymes by aspirin and other non-steroidal anti-inflammatory drugs, *J. Biol. Chem.*, 268, 6610, 1993.

7. **Barnett, J., Chow, J., Ives, D., Chiou, M., Mackenzie, R., Osen, E., Nguyen, B., Tsing, S., Bach, C., Freire, J., et al.,** Purification, characterization and selective inhibition of human prostaglandin G/H synthase 1 and 2 expressed in the baculovirus system, *Biochim. Biophys. Acta*, 1209, 130, 1994.

8. **Laneuville, O., Breuer, D. K., Dewitt, D. L., Hla, T., Funk, C. D., and Smith, W. L.,** Differential inhibition of human prostaglandin endoperoxide H synthases-1 and -2 by nonsteroidal anti-inflammatory drugs, *J. Pharmacol. Exp. Ther.*, 271, 927, 1994.

9. **Smith, W. L. and Marnett, L. J.,** Prostaglandin endoperoxide synthases, in Sigel, H. and Sigel, A., Eds., *Metal Ions in Biological Systems*, Vol. 30, Marcel Dekker, New York, 1994, 163–199.

10. **Smith, W. L. and DeWitt, D. L.,** Prostaglandin endoperoxide H synthases-1 and -2, in Dixon, F. J., Ed., *Advances in Immunology*, Vol. 62, Academic Press, San Diego, CA, 1996, 167–215.

11. **Cromlish, W. A. and Kennedy, B. P.,** Selective inhibition of cyclooxygenase-1 and -2 using intact insect cell assays, *Biochem. Pharmacol.*, 52, 1777, 1996.

12. **Wong, E., DeLuca, C., Boily, C., Charleson, S., Cromlish, W., Denis, D., Kargman, S., Kennedy, B. P., Ouellet, M., Skorey, K., O'Neill, G. P., Vickers, P. J., and Riendeau, D.,** Characterization of autocrine inducible prostaglandin H synthase-2 (PGHS-2) in human osteosarcoma cells, *Inflamm. Res.*, 46, 51, 1997.

13. **Riendeau, D., Boyce, S. P., Brideau, C., Charleson, W., Cromlish, W., Ethier, D., Evans, J., Galgueyret, J. P., Ford-Huchinson, A. W., Gordon, R., Greig, G., Gresser, M., Guay, J., Kargman, S., Leger, S., Manicini, J. A., O'Neill, G., Oullet, M., Rodger, I. W., Therien, M., Wang, Z., Webb, J. K., Wong, I., Xu, L., Young, R. N., Zamboni, R., Prasit, P., and Chan, C.C.,** Biochemical and pharmacological profile of a tetrasubstituted furanone as a highly selective cox-2 inhibitor, *Br. J. Pharmacol.*, 121, 105, 1997.

14. **DeWitt, D. L., El-Harith, E. A., Kraemer, S. A., Andrews, M. J., Yao, E. F., Armstrong, R. L., and Smith, W. L.,** The aspirin and heme-binding sites of ovine and murine prostaglandin endoperoxide synthases, *J. Biol. Chem.*, 265, 5192, 1990.

15. **Lecomte, M., Laneuville, O., Ji, C., DeWitt, D. L., and Smith, W. L.,** Acetylation of human prostaglandin endoperoxide synthase-2 (cyclooxygenase-2) by aspirin, *J. Biol. Chem.*, 269, 13207, 1994.

16. **Loll, P. J., Picot, D., and Garavito, R. M.,** The structural basis of aspirin activity inferred from the crystal structure of inactivated prostaglandin H_2 synthase, *Nature Str. Biol.*, 2, 637, 1995.

17. **The Steering Committee of the Physicians Health Study Research Group,** Preliminary report, findings from the aspirin component of the ongoing physicians health study, *N. Engl. J. Med.*, 318, 262, 1988.

18. **Lewis, H. D., Davis, J. W., Archibald, D. G., Steinke, W. E., Smitherman, T. C., Doherty, J. E., Schnaper, H. W., LeWinter, M. M., Linares, E., Pouget, J. M., Sabharwal, S. C., Chelsler, E., and DeMots, H.,** Protective effects of aspirin against acute myocardial infaction and death in men with unstable angina, *N. Engl. J. Med.*, 309, 396, 1983.

19. **Meade, T. W. and Miller, G. J.,** Combined use of aspirin and warfarin in primary prevention of ischemic heart disease in men at high risk, *Am. J. Cardiol.*, 75, 23B, 1995.

20. **Rome, L. H. and Lands, W. E. M.,** Structural requirements for time-dependent inhibition of prostaglandin biosynthesis by anti-inflammatory drugs, *Proc. Natl. Acad. Sci. U.S.A.*, 72, 4863, 1975.

21. **Kulmacz, R. J. and Lands, W. E. M.,** Stoichiometry and kinetics of the interaction of prostaglandin H synthase with anti-inflammatory agents, *J. Biol. Chem.*, 260, 12572, 1985.

22. **Callan, O. H., So, O. Y., and Swinney, D. C.,** The kinetic factors that determine the affinity and selectivity for slow binding inhibition of human prostaglandin H synthase 1 and 2 by indomethacin and flurbiprofen, *J. Biol. Chem.*, 271, 3548, 1996.

23. **Gierse, J. K., McDonald, J. J., Hauser, S. D., Rangwala, S. H., Koboldt, C. M., and Seibert, K.,** A single amino acid difference between cyclooxygenase-1 (COX-1) and -2 (COX-2) reverses the selectivity of COX-2 specific inhibitors, *J. Biol. Chem.*, 271, 15810, 1996.

24. **Copeland, R. A., Williams, J. M., Giannaras, J., Nurnberg, S., Covington, M., Pinto, D., Pick, S., and Trzaskos, J. M.,** Mechanism of selective inhibition of the inducible isoform of prostaglandin G/H synthase, *Proc. Natl. Acad. Sci. U.S.A.*, 91, 11202, 1994.

25. **Guo, Q., Wang, L. H., Ruan, K. H., and Kulmacz, R. J.,** Role of Val509 in time-dependent inhibition of human prostaglandin H synthase-2 cyclooxygenase activity by isoform-selective agents, *J. Biol. Chem.*, 271, 19134, 1996.

26. **Kargman, S., Wong, E., Greig, G. M., Falgueyret, J. P., Cromlish, W., Ethier, D., Yergey, J. A., Riendeau, D., Evans, J. F., Kennedy, B., Tagari, P., Francis, D. A., and O'Neill, G. P.,** Mechanism of selective inhibition of human prostaglandin G/H synthase-1 and -2 in intact cells, *Biochem. Pharmacol.*, 52, 1113, 1996.

27. **Anderson, G. D., Hauser, S. D., McGarity, K. L., Bremer, M. E., Isakson, P. C., and Gregory, S. A.,** Selective inhibition of cyclooxygenase (COX)-2 reverses inflammation and expression of COX-2 and interleukin 6 in rat adjuvant arthritis, *J. Clin. Invest.*, 97, 2672, 1996.

28. **Chan, C. C., Boyce, S., Brideau, C., Ford-Hutchinson, A. W., Gordon, R., Guay, D., Hill, R. G., Li, C. S., Mancini, J., Penneton, M., et al.,** Pharmacology of a selective cyclooxygenase-2 inhibitor, L-745,337, a novel nonsteroidal anti-inflammatory agent with an ulcerogenic sparing effect in rat and nonhuman primate stomach, *J. Pharmacol. Exp. Ther.*, 274, 1531, 1995.

29. **Futaki, N., Takahashi, S., Yokoyama, M., Arai, I., Higuchi, S., and Otomo, S.,** NS-398, a new anti-inflammatory agent, selectively inhibits prostaglandin G/H synthase/cyclooxygenase (COX-2) activity *in vitro*, *Prostaglandins*, 47, 55, 1994.

30. **Glaser, K.B.,** Cyclooxygenase selectivity and NSAIDs, cyclooxygenase-2 selectivity of etodolac (lodine), *Inflammopharmacology,* 3, 335, 1995.
31. **Glaser, K., Sung, M. L., O'Neill, K., Belfast, M., Hartman, D., Carlson, R., Kreft, A., Kubrak, D., Hsiao, C. L., and Weichman, B.,** Etodolac selectively inhibits human prostaglandin G/H synthase 2 (PGHS-2) versus human PGHS-1, *Eur. J. Pharmacol.,* 281, 107, 1995.

Chapter

4

Phospholipid Growth Factors: Identification and Mechanism of Action

Gábor J. Tigyi, Károly Liliom, David J. Fischer,
and Zhong Guo

Contents

I. Introduction

Nontransformed cells depend on growth factors for their survival, proliferation, and differentiation. In addition to polypeptide growth factors, an emerging group of naturally occurring phospholipid growth factors (PLGFs) has been discovered (for reviews, see References 37 and 57). Lysophosphatidic acid (1-acyl-2-lyso-sn-glycero-3-phosphate, LPA), cyclic-phosphatidic acid (cyclic-PA), plasmalogen-glycerophosphate (alkenyl-GP), phosphatidic acid (PA), lysophosphatidylserine (LPS), sphingosine-1-phosphate (SPP), and sphingosylphosphorylcholine (lysosphingomyelin, SPC) are all naturally occurring lipid mediators with growth factor–like actions. The best-characterized member of this group is LPA, nature's simplest phospholipid. LPA elicits responses in almost every cell type spanning the phylogenetic tree, from *Dictyostelium* to humans. These biological effects include (1) the mitogenic or antimitogenic regulation of the cell cycle[50,61]; (2) regulation of Ca^{2+} homeostasis that leads to the contraction of smooth muscle cells,[58,59] the induction of inward currents in oocytes,[11,15,16,55] and neurotransmitter release[45]; (3) regulation of the actin cytoskeleton, affecting cell shape,[44,55] migration,[27,67] and tumor cell invasiveness[25]; and (4) the inhibition of apoptosis[60,63] and differentiation.[12,26,51,52] PLGFs elicit cellular responses via multiple G protein–coupled receptors, which have distinct pharmacological and signal transduction properties.[1,14,21,22] LPA is generated from activated and injured cells,[18] including platelets[13,19,28,29]; blood serum is consequently a very rich source of LPA.[44,49,53] In addition to LPA, serum contains at least eight other lipids that constitute ~90% of the LPA-like biological activity.[55]

In this chapter, a survey of the techniques frequently used in the analysis of PLGFs is provided. Our goal is to describe well-tested protocols for the reader interested in exploring PLGFs and their receptors. Accordingly, lipid biochemical techniques applicable to the purification and identification of PLGFs are first described, followed by the description of cellular assays for their monitoring, and finally, techniques used for the analysis of PLGF receptors (PLGFRs).

II. Protocol

A. Isolation and Purification

1. Extraction of PLGFs from biological fluids

Biological fluids are prepared for extraction by freeze-drying to remove the water. The extraction of PLGFs is carried out in a two-step procedure. First, the freeze-dried material is extracted with a mixture of chloroform and methanol at a ratio of 2:1 (v/v) containing 50 μg/ml butylated hydroxytoluene antioxidant (BHT, Cat. No. B-1378, Sigma Chemical Co., St. Louis, MO). Second, the residual dry material is extracted with methanol containing 50 μg/ml BHT.

Solvents (per 1 g of lyophilized material):

a.	Chloroform	20 ml
	Methanol	10 ml
	BHT	1.5 mg
b.	Methanol	30 ml
	BHT	1.5 mg

Procedure

1. Add 5× volume (v/w) of the chloroform/methanol/BHT mixture to the lyophilized material and shake vigorously for 10 min using a rotary shaker.

2. Centrifuge at $700 \times g$ for 10 min and collect the supernatant.

3. Repeat steps 1 and 2 two more times.

4. Add a 5× volume of methanol/BHT to the last pellet and shake vigorously for 10 min.

5. Centrifuge at $700 \times g$ for 10 min and collect the supernatant.

6. Repeat steps 4 and 5 two more times.

7. Pool the supernatants from the chloroform/methanol/ BHT extraction and the methanol/ BHT extraction separately. Dry the extracts under nitrogen at 40°C (e.g., Reacti-Vap evaporator attached to a Reacti-Therm heating module, Pierce Chemical Co., Rockford, IL) in preweighed vials. When large amounts of lyophilized material are being processed, extracts can be dried using a rotary evaporator.

8. Weigh the dried extract, seal it under argon in glass vials, and store at −70°C.

For rabbit and human serum, the chloroform/methanol/BHT extract constitutes ~6% of the wet weight, whereas the methanol-soluble material comprises ~3%. This method is suitable for extracting polar lipids from a wide variety of biological fluids. For a comprehensive description of lipid extraction methods, the reader is referred to a paper by Radin.[43]

2. Purification of PLGFs by HPLC

Analytical scale separation of PLGFs and related lipids present in the methanol extracts of biological fluids can be done by high-performance liquid chromatography (HPLC).

1. *Column:* Microsorb Dynamax (Rainin Instruments, Woburn, MA; Cat. No. 80-125-C5), 4.6 × 250 mm, packed with 5 μm spherical silica with 100 Å pore size. Protect the column using a 4.6 × 15 mm guard column with the same packing material (Cat. No. 80-100-G5).

2. *Solvents:*

Eluent A

Chloroform	600 ml
Methanol	350 ml
Deionized HPLC-grade water	45 ml
Ammonium hydroxide (30%)	5 ml

Eluent B

Chloroform	300 ml
Methanol	600 ml
Deionized HPLC-grade water	95 ml
Ammonium hydroxide (30%)	5 ml

3. *Gradient elution program:* The column should be equilibrated with 10 column volumes of solvent A at a flow rate of 1 ml/min. A linear gradient is then established by increasing the amount of eluent B to 100% over 25 min, followed by an isocratic application of eluent B for 30 min.

Solvents should be prepared fresh and, after mixing, filtered through an 0.2-µm Teflon filter. The gradient is formed by a two-pump solvent delivery system (Waters, Milford, MA, controlled by the Millennium 2010 v2.1 software package).

The ultraviolet (UV) absorption coefficient of simple phospholipids is low, which is complicated in the far-UV range by the high absorption of organic solvents often used in HPLC. To overcome the difficulty of low UV absorption, the use of an evaporative light scattering detector (ELSD) is advised. The light-scattering signal is proportional to the mass of the eluting solute, providing a sensitive detection of compounds, independent of their chemical nature.[5] With the use of a metering valve, it is possible to split the column effluent between the detector and a fraction collector. In the authors' HPLC system, an ELSD (Varex Model II A, Varex Corp., Burtonsville, MD) is used. To monitor the effluent and collect fractions for further analysis, a metering valve splits the effluent at a 1:4 ratio between the detector and the fraction collector. The retention times of known PLGFs and of some related phospholipids are given in Table 1.

3. Analysis of PLGFs by thin-layer chromatography

The high resolving power of two-dimensional thin-layer chromatography (2D-TLC) makes it an ideal analytical tool for the rapid characterization and identification of PLGFs. 2D-TLC of PLGFs is carried out on 20 × 20 cm silica plates with a thickness of 250 µm (e.g., Whatman K6, 60 Å pore size, Cat. No. 4860-820, Whatman, Inc., Fairfield, NJ). Plates must be activated before use by baking at 100°C for at least 30 min. Glass chromatographic chambers (Cat. No. K416180-0000, Fisher Scientific, Fairlawn, NJ) are lined with filter paper to provide a saturated solvent atmosphere.

1. *Solvent A* (first dimension)

Chloroform	120 ml
Methanol	80 ml
Ammonium hydroxide (30%)	20 ml

TABLE 1
Characterization of PLGFs and Related Phospholipids by HPLC Retention Times (Rt) and 2D-TLC Retention Factors (Rf I and Rf II)

Compound	Rt (min)	Rf I	Rf II
Alkenyl-glycerophosphate	12	0.12	0.74
Cyclic-phosphatidic acid	6	0.70	0.86
Lysophosphatidic acid	16	0.06	0.60
Lysophosphatidylcholine	30	0.20	0.11
Lysophosphatidylethanolamine	10	0.36	0.49
Lysophosphatidylglycerol	6	0.35	0.58
Lysophosphatidylserine	17	0.26	0.24
Lyso-platelet-activating factor	29	0.21	0.10
N-Palmitoyl-serine phosphoric acid	27	0.02	0.17
N-Palmitoyl-tyrosine phosphoric acid	24	0.03	0.29
Phosphatidic acid	14	0.15	0.84
Platelet-activating factor	19	0.17	0.08
Sphingomyelin	15	0.22	0.44
Sphingosine-1-phosphate	22	0.05	0.50
Sphingosylphosphorylcholine	35	0.13	0.04

For solvents and chromatography conditions, see text.

2. *Solvent B* (second dimension)

Chloroform	60 ml
Acetone	80 ml
Methanol	20 ml
Acetic acid (glacial)	20 ml
Deionized water	10 ml

The lipid mixture is deposited on the activated plate with a glass capillary tube or with a Hamilton syringe, 2 cm from the corner; up to 4 mg of material can be spotted. After development in the first dimension, the plate is rapidly dried with a hair dryer, and rotated 90° for development in the second dimension. Following development in the second dimension, lipids are visualized with primulin staining and observed under UV illumination.

3. *Primulin staining:*
   a. Spray the TLC plate evenly with ~10 ml dilute primulin.
   b. After marking the fluorescent lipid spots with a soft pencil, either scrape the lipids off the plate or stain them further with reagents that are specific for particular chemical groups of interest.
4. *Free amine identification:* Free amines, like those present in sphingoid bases, can be visualized with ninhydrin reagent (Cat. No. N-0507, Sigma).

5. *Phosphate moiety detection:* Phosphate moieties can be detected with the molybdenum blue spray reagent.

6. When spots are removed for testing biological activity, or for mass spectrometric analysis, PLGFs can be extracted from the silica with three washes of excess methanol (10×, v/w). The silica suspension is sedimented by centrifugation, and the methanol supernatant is filtered through an 0.2-μm Teflon or nylon syringe filter before drying under a stream of nitrogen at 40°C.

Reliable identification of PLGFs can be achieved in a 2D-TLC system based on matching retention factors (Rf) with those of known standards.[33,55] The Rf values of known phospholipid standards are shown in Table 1. Figure 1 shows the separation of PLGFs in a methanol extract of human serum albumin.

B.   Monitoring PLGF Activity

1.   The *Xenopus* oocyte bioassay

The *Xenopus* oocyte, largely because of its size of ~1.5 mm in diameter, has become an invaluable tool in studying receptor-activated ion currents and signal transduction mechanisms. The ovarian follicle, which is an electrically coupled syncytium between follicular cells covering the oocyte and the oocyte proper, endogenously expresses many different types of receptors (see, for a review, Miledi et al.[35]) for hormones, e.g., gonadotropins, vasoactive intestinal peptide (VIP), oxytocin; neurotransmitters, e.g., acetylcholine, ATP, adenosine; and mediators e.g., substance P, atrial natriuretic peptide (ANP). Receptors for lipid mediators are among these, which also include prostaglandin receptors,[36] and multiple subtypes of PLG-FRs.[11,15,16,34,49,53,55] LPA has been identified as a ligand that activates oscillatory Cl⁻ currents via the G protein–mediated production of inositol 1,3,4-trisphosphate, causing the release of Ca^{2+} from intracellular stores, which in turn opens Ca^{2+}-activated Cl⁻ channels (Figure 2).[40,41,49,55] Perhaps the biggest advantage of the oocyte bioassay is its high sensitivity to PLGFs, as LPA and alkenyl-GP elicit oscillatory Cl⁻ currents at nanomolar concentrations (Figure 3).

Xenopus oocytes have been used in studies of PLGFs that include (1) the detection and quantification of PLGFs in biological fluids[32,33,49,53–55,66]; (2) pharmacological analysis of PLGF-agonists and antagonists[2,23,31,34,56]; and (3) genetic and pharmacological analysis of different PLGFR.[17,21,31,34]

Many analogs of LPA elicit oscillatory Cl⁻ currents in the oocytes through multiple subtypes of the PLGFR expressed simultaneously.[17,21,34] LPA is a promiscuous ligand that activates at least three pharmacologically different receptor subtypes expressed in the oocyte.[17,21,34] Therefore, a thorough pharmacological analysis must always be performed when assessing LPA-like ligands in the oocyte bioassay. SPP and SPC do not elicit oscillatory Cl⁻ currents in the oocyte; for these lipids, other bioassays are available.[3,4] Commercially available SPP and SPC obtained from certain sources (e.g., Sigma Chemical Co., Toronto Biochemicals, Inc.) often contain an alkenyl-GP impurity[17,32] that can lead to ambiguous results. In the oocyte assay, the time between application of the test compound and current

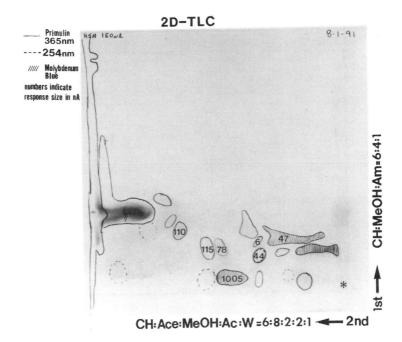

FIGURE 1

Two-dimensional TLC separation of human serum albumin-bound lipids reveals multiple PLGFs. A total of 2 mg of the HSA-bound lipids was separated and stained with primulin, and molybdenum blue. The fluorescence quenching of the indicator included in the silica gel plate (250-μm thickness, 20 × 20 cm) caused by the lipids was visualized under UV illumination at a wavelength of 254 nm (dashed line). The lipid spots that stained with primulin or caused fluorescence quenching were scraped off and tested in the oocyte bioassay at 250-fold dilution. The numbers in the spots indicate the peak current amplitude of the oscillatory Cl⁻ current elicited by each lipid. The spot that gave 1005 nA current had a mobility identical to the LPA standard.

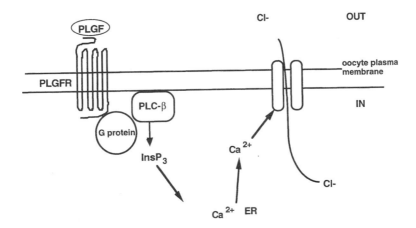

FIGURE 2

The mechanism of the oscillatory Cl⁻ currents elicited by PLGFs in the oocyte.

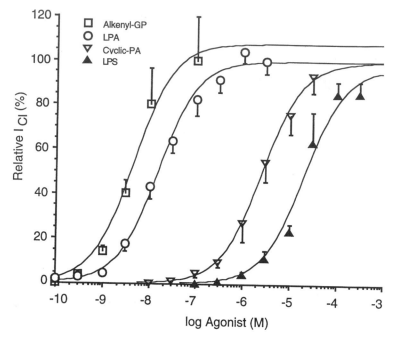

FIGURE 3

Dose–response relationship of the oscillatory Cl⁻ current elicited by different PLGFs. Alkenyl-GP (squares), LPA (circles), cyclic-PA (open triangles), and LPS (filled triangles) all elicit oscillatory Cl⁻ currents with varying potencies, as indicated by the different apparent EC_{50} concentrations.

response is on the order of seconds, diminishing the chance of metabolic inter-conversion. In contrast, to elicit a response, cell proliferation assays require the presence of lipids for several hours, during which metabolic interconversion (e.g., PA into LPA) could occur. However, the oocyte assay has its limitations. The oocyte expresses several, but not all, subtypes of PLGFRs. For example, none of the Edg family of PLGFR is expressed in the oocyte (see below), and, when expressed heterologously, these receptors do not elicit oscillatory Cl⁻ currents, presumably because of the lack of coupling between these receptors and the oocyte signal transduction systems (Virag, Guo, and Tigyi, unpublished results). Conse-quently, the oocyte bioassay does not replace the need for other cellular assay systems in the assessment of PLGFs.

a. Isolation and Preparation of Oocytes for Voltage-Clamp Recording

There are several book chapters on the use of *Xenopus* oocytes for the electrophysio-logical recording of endogenous, as well as heterologously expressed receptors.[7,8,46,48] The reader referred to these publications for a detailed description of the equipment and principles of the voltage-clamp recording technique. Here, we describe only protocols for the isolation of oocytes and some of the procedures followed during voltage-clamp recording.

Removal of frog oocytes

1. Anesthetize the frog with 0.2% tricaine (3-aminobenzoic acid ethylester, Sigma, A-5040) in tap water until the righting reflex is abolished. Lay the anesthetized frog on its back on a sterile surgical drape under sufficient lighting.

2. Observe sterile conditions. Make an ~1-cm parasagittal incision in the lower abdomen. Because the skin is extremely tough, pierce it with a sterile needle and start the incision from this hole with sharp scissors. After cutting the skin, cut through the abdominal muscle layer and pull out the desired number of lobes of the ovary. Tie down the ovary with a surgical suture and cut it below the knot. Suture the wound layer-by-layer, using a silk surgical suture for the muscles and fascias (e.g., Ethicon K-871H, two to three stitches per layer) and a Mersilene suture (e.g., Ethicon R633H) for the skin. Allow the frog to recover in shallow tap water. Make sure that the nostrils are above the water by laying the head on a piece of sponge. The entire operation should not take more than 10 min, and the frog should recover within 30 min from the anesthesia.

3. Rinse the ovary in Ca^{2+}-free Ovarian Ringer's-2 (OR-2) solution thoroughly two to three times. Separate the individual lobes and open them up by cutting along the perimeter of the lobe. Rinse the open lobes two more times in OR-2 solution in separate petri dishes.

4. Isolate oocytes by collagenase treatment. For this, use Type A collagenase from Boehringer Mannheim (Cat. No. 1088 785) at 1.4 mg/ml (~0.5 U/ml) in a Ca^{2+}-free OR-2 solution. Place each half lobe into 1 ml of collagenase in a small screw-cap glass vial (e.g., Solvent Saver Scintillation Vials, Kimble, Inc.) and place on a rotator (e.g., Pellco 2, Model No. 1050, Ted Pella, Inc., Redding, CA) for 30 min at room temperature. After 30 min, gently vortex the vial and look for free-floating oocytes in the solution. Continue the treatment until approximately one half of the oocytes are freed from the ovary (in general, do not exceed a 1-h treatment); then stop the treatment by removing the collagenase solution and replacing it with ~5 ml of defolliculation solution. Gentle vortexing will detach the enveloping cells, leading to the appearance of floating "ghosts" that are the follicular envelopes. If the envelopes do not detach with shaking, they can be manually peeled off using a watchmaker's forceps (Dumont No. 5) under a dissecting microscope. The defolliculated oocytes should be transferred to Barth's solution, using a Pasteur pipette with a widened opening that will not damage the oocytes, and rinsed two times.

5. Keep oocytes in Barth's solution in an incubator between 17 and 20°C and use them for electrophysiological recording 1 to 7 days after isolation. When using the oocytes for the heterologous expression of the PSP24 PLGFR, inject collagenase-treated oocytes with 0.2 ng cRNA in 25 nl diethyl pyrocarbonate-treated double-distilled water. Because of the rapid expression of the receptor mRNA, electrophysiological recording is done 2 to 4 days after injection, with the biggest responses usually recorded on days 2 and 3. The amount of mRNA and the optimal recording times should be determined empirically for every receptor.

The authors generally use stage V[10] (1.0 to 1.2 mm in diameter, with differentiated hemispheres and a dark animal pole) and VI oocytes (>1.2 mm in diameter, distinguished by an unpigmented equatorial band). However, oocytes at earlier stages of maturation (stage II or greater) are also responsive to LPA (Tigyi, unpublished). Oocytes are checked under a dissecting microscope daily. Damaged cells, showing a breakdown of pigmentation, are discarded to prevent spoilage of the neighboring oocytes, and the Barth's medium is replaced daily.

b. Voltage-Clamp Recording for PLGF-Elicited Membrane Currents

Voltage-clamp recording is the simplest and most straightforward electrophysiological technique for monitoring PLGF-elicited membrane currents in oocytes. With a limit of detection of approximately 1 nA, it is possible to measure the opening of ~1000 channels with each contributing 1 pA of current, making the assay highly sensitive. There are commercially available voltage-clamp amplifiers especially designed for oocytes that can produce large output currents in the 10-μA range (Axon, Inc., Warner Instruments, Inc., Dagan Instruments, Inc.). For building and operating a voltage clamp setup, the reader is referred to other publications.[47,48]

The authors routinely use a perfusion chamber that is connected to reservoirs (test tubes for the samples) fed by gravitation through plastic tubing (Figure 4) at a flow rate of 1 to 5 ml/min. Recording electrodes are pulled from glass capillaries with an internal filament and are filled with 3 M KCl. The current electrode used for clamping large currents should have a low resistance of a few MΩ. To achieve this, the tip of the electrode is broken by pressing the two electrodes against each other or against the bottom of the recording chamber. The low resistance of the recording electrode diminishes the noise in the system. Healthy oocytes should have a resting membrane potential of less than –50 mV and an input resistance of 1 MΩ or more. For recording PLGF-elicited oscillatory currents, the membrane potential is held at –60 mV, which is sufficiently far from the Cl⁻ equilibrium potential, which is around –24 mV in the oocyte. The oocyte, after its membrane is pierced with the electrodes and its membrane potential is clamped at –60 mV, is allowed to stabilize for 30 min. This time is necessary to allow the membrane to seal and to allow the removal of the Ca^{2+} that has entered the cytoplasm when the membrane is pierced.

PLGFs can be applied through superfusion or by focal ejection from a micropipette brought into close proximity to the oocyte surface. For focal ejection onto the oocyte surface, as well as for the intracellular delivery of compounds, a pressure microinjector (e.g., Pneumatic PicoPump, Model PV 800, World Precision Instruments, Sarasota, FL) is used. However, superfusion with frog Ringer's solution is routinely used, and the test lipids, or biological fluids, are diluted in it. Because the Ca^{2+} salts of the acidic phospholipids are insoluble in aqueous solvents and are consequently inactive in the bioassay, care must be taken for their proper dissolution. For LPA and other PLGFs, the following dissolution procedure is used:

1. 100 μl of 10 mM LPA in methanol (stabile for 1 month at –70°C);
2. 10 μl of 100 mM ethylene glycol-bis(β-aminoethyl ether)-tetraacetic acid (EGTA), pH 7.4;
3. 890 μl 1 mM bovine serum albumin (BSA), fatty acid–free in Ca^{2+}-free Hanks' solution (Gibco-BRL, Gaithersburg, MD), pH 7.4, with NaOH, sterile filtered.

This stock solution is sonicated in a bath sonicator for ~10 to 20 s and is stabile for 1 week if kept on ice during daily use.

LPA is sticky, and nanomolar–micromolar working dilutions lose their activity rapidly (>30 min). Therefore, working dilutions should always be freshly prepared

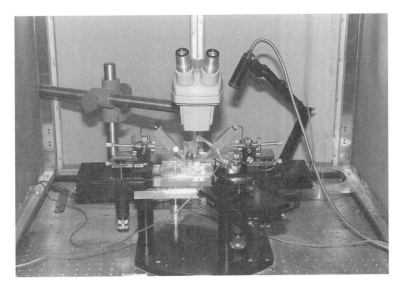

FIGURE 4
A panoramic view of the two-electrode voltage-clamp recording setup and perfusion chamber used for *Xenopus* oocytes.

immediately before application. Final working dilutions of the albumin-complexed PLGFs are made in frog Ringer's solution or in the appropriate culture medium when mammalian cells are to be tested. When a dilution lower than 100-fold is to be used, methanol (1%) alone must be tested because a concentration $\geq 1\%$ itself can elicit oscillatory currents in oocytes of some frogs (Tigyi, unpublished) and alcohols have been shown to potentiate oscillatory currents.[24]

The peak amplitude, as well as the area under the response trace, elicited by LPA and other PLGFs shows a sigmoidal dose–response relationship (see Figure 3). This allows the titration of an unknown PLGF activity present in a biological fluid and the estimation of its concentration expressed in the equivalent concentration of LPA that elicits the same response.[33,54,66] To keep receptor desensitization to a minimum, dilute the test samples so that they give currents comparable to 5 to 10 nM of LPA (oleoyl; typically not more than 100 to 200 nA), and always wait 15 min before the next application to allow for appropriate washout and recovery from desensitization. When screening multiple specimens for PLGFs, also include a 5 to 10 nM test concentration of LPA after every third unknown sample, which typically gives a current that is within $\pm 10\%$ of the size of previous LPA response.

2. PLGF-elicited calcium responses in mammalian cells

With the use of the fluorescent Ca^{2+} indicators fura-2 and indo-1 (Molecular Probes, Inc., Eugene, OR), changes in intracellular Ca^{2+} concentrations can be used to monitor the activation of G protein–coupled receptors by PLGFs. The release of Ca^{2+} from internal stores is detected by changes in the fluorescence of the Ca^{2+} indicators. For use in ratioimaging, fura-2 has become the indicator of choice. Upon binding Ca^{2+}, fura-2 undergoes a shift in absorbance, which is reflected in a shift in its excitation spectrum. This shift can be monitored by measuring its emission at 510 nm using excitation wavelengths of 340 and 380 nm. The resulting change in emission is reported as a ratio of $A_{340/380}$, which is monitored in a dual-wavelength spectrofluorometer.

PLGFs elicit a variety of cellular responses in fibroblast cells; consequently, these cells (NIH3T3, Swiss 3T3, Rat 1, TIG-3) are commonly used to monitor PLGF activity in various assays. To monitor the fura-2 fluorescence shift associated with changes in intracellular Ca^{2+} concentrations, $[Ca^{2+}]_i$, a Perkin Elmer spectrofluorometer (model LS-50) fitted with a water-jacketed cuvette holder to maintain the cells at 37°C is used, along with the ICBC (Perkin-Elmer) software package provided with the instrument.

Plating Fibroblasts for Fura-2 Measurements

1. Prepare glass coverslips (9 × 35 mm) by washing them with concentrated HNO_3 and thoroughly rinsing with distilled water. Place coverslips in corresponding Leighton culture tubes (Bellco Glass, Inc., Vineland, NJ) and sterilize.

2. Plate 2 to 3×10^5 NIH3T3 fibroblasts onto coverslips in growth medium (Dulbecco's Modified Minimal Essential Medium, or DMEM, + 10% FBS).

3. When the cells have reached approximately 70 to 80% confluence, change the medium to DMEM alone for a minimum of 12 to 18 h (usually overnight, depending on the sensitivity of the cells to the serum withdrawal).

Cell Loading with Fura 2-AM

1. Prior to loading the cells with fura-2 AM (cell-permeant form of acetoxy methylester), measure the autofluorescence of the cells. Record the emission at 510 nm using excitation wavelengths of 340 and 380 nm.

2. Dissolve fura-2 AM (Molecular Probes, Cat. No. F-1221) in DMSO at a concentration of 1 mM. Dilute the fura-2 AM to a final concentration of 5 μM, along with Pluronic acid F-127 (Molecular Probes, Cat. No. P-1572) to a concentration of 0.01%, directly in the culture media.

3. Load the cells for 15 to 20 min at 37°C in the dark.

4. Wash the cells with Krebs buffer and incubate for 10 min at 37°C.

Monitoring Changes in $[Ca^{2+}]_i$

1. Transfer the coverslip to a cuvette contained in the spectrofluorometer and perfuse with Krebs solution, prewarmed to 37°C, at a flow rate of 3 ml/min.

2. Measure the fluorescence emission at 510 nm using excitation wavelengths of 340 and 380 nm. Compare the values to autofluorescence values to assess the loading efficiency.

3. Establish a steady baseline and begin the experimental protocol. Test PLGFs by perfusing them into the cuvette at the same rate.

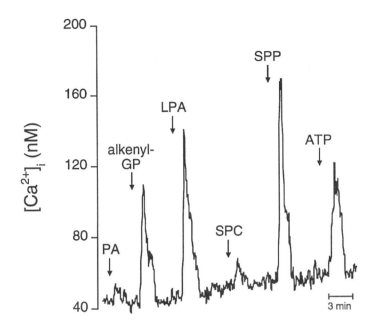

FIGURE 5

Ca^{2+} transients elicted by PLGFs in NIH3T3 cells. Cells were serum starved in DMEM for 4 h, loaded with 5 μM fura-2 AM for 20 min, and treated with 1 μM PA, alkenyl-GP, LPA, SPC, and ATP, and 100 nM SPP.

4. At the end of the experiment, calibrate the fluorescent signal by measuring the maximum fluorescence (R_{max}) in the presence of 4 mM CaCl$_2$ and 5 μM ionomycin, and the minimum fluorescence (R_{min}) in the presence of 4 mM EGTA and 5 μM ionomycin.

5. Calculate [Ca^{2+}]$_i$ with the ICBC software.

The ability of the cells to load efficiently with fura-2 is dependent on the cell type and loading conditions. In general, most cells can be sufficiently loaded using a fura-2 AM concentration between 1 and 5 μM, for 15 to 60 min. It is recommended to use the lowest concentration possible to avoid problems associated with the aldehyde by-products of the acetoxy methylester hydrolysis. For individual cell types, specific loading conditions should be empirically determined.

Intracellular Ca^{2+} concentrations are calculated using the ICBC software according to the equation $[Ca^{2+}] = K_d(R - R_{min}/R_{max} - R)(S_{f2}/S_{b2})$,[20] where K_d is the Ca^{2+} dissociation constant for fura-2, R is the fluorescence ratio (340/380), R_{min} and R_{max} are the minimum and maximum fluorescent ratios, respectively, and S_{f2}/S_{b2} is the ratio of R_{min} to R_{max} at 380 nm.[20] The K_d for fura-2 is dependent on pH, temperature, and ionic strength.[30] Since these parameters vary from cell to cell, the K_d value inside one cell type may not apply to another. Therefore, to avoid inaccurate estimations, the cell type dependence of the K_d value must be taken into consideration when calibrating intracellular Ca^{2+} concentration. A representative trace of PLGF-elicited Ca^{2+} transients is shown in Figure 5.

3.   Cell proliferation assays

The effect of PLGFs on the cell cycle is monitored by measuring cell proliferation through the incorporation of [3H]thymidine into DNA, or by direct cell counting. Fibroblasts are the cell of choice for measuring the mitogenic activity,[9,50,62] under defined, serum-free conditions.

Plating and Cell Treatment

1.   Plate Swiss 3T3 cells in 24-well plates at a density of 1.5×10^4/cm^2 in DMEM containing 10% calf serum.

2.   Feed the cells 2 days later and grow to confluence. Use the cells 3 to 4 days after plating, when they are confluent and quiescent.

3.   Wash the cells with DMEM and add 1 ml of DMEM:Waymouth medium (1:1) containing 20 µg/ml BSA (fatty acid–free, Sigma, Cat. No. A-6003) and 5 µg/ml transferrin (Collaborative Biomedical Products, Cat. No. 40204) to each well.

4.   Add the PLGFs, diluted to the appropriate concentration in fatty acid–free BSA (dissolved in Hanks'-BSA solution as described under the oocyte bioassay).

5.   After 18 h, add 1 µCi of [^{3}H]thymidine (5.0 Ci/mmol, Amersham) and incubate for 6 h.

Measuring [^{3}H]Thymidine Incorporated into DNA

1.   Aspirate the medium from the cells using a 24-well cell harvester (Brandel, Gaithersburg, MD), through an FP-100 Whatman GF/B filter.

2.   Rinse with PBS, followed by a rinse with Ca^{2+}-free Hanks' solution.

3.   Add 0.5 ml of 0.05% trypsin in 0.53 mM EDTA, pH 7.4, to each well and incubate until the cells begin to detach from the well (5 to 10 min).

4.   Triturate each well to ensure that all cells are detached.

5.   Aspirate cells, and rinse with 0.14 mM PBS.

6.   Rinse each well two times with 7.5% trichloroacetic acid (w/v) and one time with PBS.

7.   Remove the individual filters from the cell harvester, place in 5 ml of scintillation cocktail, and quantitate the acid-insoluble material by liquid scintillation counting.

Some PLGFs, cyclic-PA and LPS in particular, have been shown to be antimitogenic.[38,50,64,65] To measure the antimitogenic activity, cells must be stimulated to proliferate to show an inhibitory effect on cell growth. For this, subconfluent cultures of NIH3T3 cells are used under defined, serum-free conditions. While this assay serves as a good method for monitoring antimitogenic activity, it can also be used to measure the mitogenic activity of PLGFs.

Plating and Cell Treatment

1.   Plate NIH3T3 cells in 24-well plates at a density of 1×10^4/cm^2, in 2% FBS, and culture for 24 h.

2. Rinse the cells with PBS and change the medium to MCDB-104 containing 25 ng/ml basic fibroblast growth factor, 5 µg/ml insulin, 5 µg/ml transferrin (all from Collaborative Biomedical Products), and 10 ng/ml dexamethasone.

3. Add the PLGFs, diluted to the appropriate concentration in fatty acid–free BSA (10 µM final concentration added in a total volume of 20 µl).

Cell Counting

1. After 3 days, remove the medium from the cells and transfer it to a counting vial.

2. Rinse the cells one time with 1 ml of Ca^{2+}-free Hanks'; transfer the supernatant to the counting vial, and add 1 ml of 0.05% trypsin in 0.53 mM EDTA, pH 7.4, to each well.

3. When the cells begin to detach from the well (after about 5 min), triturate to remove all the cells, and transfer to the vial.

4. Add 7 ml of Isoton II (Coulter Corporation, Hileah, FL) to each vial, mix, and count with a Coulter counter.

C. Analysis of PLGF Receptors

Most cell types are responsive to PLGFs. Consequently, the lack of cell types devoid of PLGFRs has been a major setback. Recently, *Spodoptera frugiperda* insect cells[39,68] and yeast cells[14] have been used for the functional expression of PLGFRs. However, both of these cell types have serious limitations. Insect cells lack the $G_{\alpha i2}$ heterotrimeric G protein,[42] which couples PLGF activation to the activation of the mitogen-activated protein (MAP) kinase cascade.[6] Not every PLGFR couples to the yeast heterotrimeric G protein, and even those that do couple require high micromolar ligand concentrations.[14] Heterologous expression of PLGFR is also complcated by the fact that they signal through the non-ligand-mediated transactivation of tyrosine kinase receptors (e.g. EGF or PDGF receptors). Thus, the simultaneous expression of these tyrosine kinases is required for the reconstitution of some PLGF effects in a heterologous system. Liver tissue shows minimal expression of PLGFR mRNA. The Hep G2 human hepatoma cell line was found to be nonresponsive to LPA (Dr. Kees Jalink, personal communication) and devoid of any of the PSP24 or Edg receptors (Figure 6). Establishing Hep G2 clones that stably express the PSP24 and Edg receptors has provided a very useful assay for the characterization of the pharmacological and signal transduction properties of these receptors.

1. Screening for PLGFRs by RT-PCR

a. Isolation of mRNA

Extreme care must be taken to ensure that this step is carried out under RNase-free conditions.

Edg1   Edg2   Edg3   Edg5   PSP24

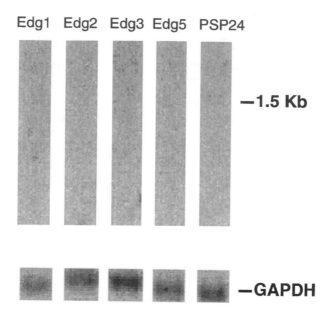

FIGURE 6

Northern blot analysis of the human hepatoma cell line Hep G2 shows no expression of mRNA for the PLGFRs in the PSP and Edg receptor families; 15 μg total RNA was loaded and hybridized with probes containing the full coding sequence of the different Edg family and the PSP24 receptors. The membranes were also hybridized with the glyceraldehyde 3-phosphate dehydrogenase (GAPDH) probe as a positive control.

1. Homogenize fresh or frozen tissue (e.g., Polytron, 15 to 20 s) in TRIzol reagent (1:9, v/v) (Gibco-BRL Technologies).

2. Incubate the homogenized samples for 5 min at room temperature. Add 0.2 ml of chloroform per 1 ml of TRIzol reagent. Shake tubes vigorously by hand for 15 s and allow the phases to separate for 2 to 3 min. Centrifuge the samples at $12,000 \times g$ for 15 min in a refrigerated centrifuge at 4°C.

3. Transfer the upper aqueous phase to a fresh (RNase-free) tube, add 0.5 ml of isopropyl alcohol per 1 ml of TRIzol reagent, and mix thoroughly. Incubate samples at room temperature for 10 min and centrifuge at $12,000 \times g$ for 10 min at 4°C.

4. Carefully aspirate the supernatant and wash the RNA pellet with 1 ml of 75% ethanol. Mix the sample by vortexing and centrifuge at $7500 \times g$ for 5 min at 4°C.

5. Remove the supernatant and dry the sample in a Speed Vac for no more than 5 to 10 min. Dissolve RNA in RNase-free water by vortexing and incubation at 55 to 60°C for 10 min.

b. Primer Design

The primers used in the reverse transcriptase-polymerase chain reaction (RT-PCR) are designed from the cDNA sequences of the PSP and EDG families of PLGFRs

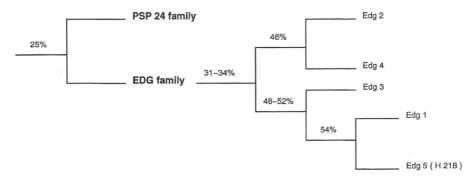

FIGURE 7
Amino acid sequence identities within the PLGFRs. Numbers in parenthesis indicate percent identities between the individual receptors.

(Figure 7). All primers contain a Bgl II restriction site in the 5′-forward primer and a Not I restriction site in the 3′-reverse primer. These two unique restriction sites enable the directional cloning of the PCR products into many mammalian expression vectors, including the pcDNA3.1 vector (Invitrogen, Inc., Carlsbad, CA).

PSP 24 (Genbank accession No. U92642)
    Forward primer: AAAAGATCT$\underline{A_1T}$GGCCCTGCAACAGCAC
    Reverse primer: AAAGCGGCCGC$\underline{CTA}_{1130}$AACCGCAGACTGGTT
Edg-1 (Genbank accession No. M31210)
    Forward primer: AAAAGATCTCC$\underline{A_1T}$GGGGCCCACCAGC
    Reverse primer: AAAGCGGCCGC$\underline{CTA}_{1146}$GGAAGAAGAGTT
Edg-2 (Genbank accession No. U80811)
    Forward primer: AAAAGATCTTC$\underline{A_1T}$GGCTGCCATCTCT
    Reverse primer: AAAGCGGCCGC$\underline{CTA}_{1095}$AACCACAGAGTG
Edg-3 (Genbank accession No. X83864)
    Forward primer: AAAAGATCTGCCAAGTG$\underline{A_1T}$GGCA
    Reverse primer: AAAGCGGCCGCGCGACGCATGCAATA$_{1178}$
Edg-4 (Genbank accession No. AF011466)
    Forward primer: AAAAGATCTGCCCAG$\underline{A_1T}$GGTCATC
    Reverse primer: AAAGCGGCCGCCAG$\underline{TCA}_{1148}$GTCAGTCCTGTTGG
Edg-5 (also known as H218, Genbank accession No. AF022138)
    Forward primer: AAAAGATCT$\underline{A_1T}$GGGCGGTTTATACTC
    Reverse primer: AAAGCGGCCGC$\underline{TCA}_{1191}$GACCACTGTGTTGC

Note: *Underlined nucleotides represent the start and stop codons of the open*
           reading frame.

c. RT-PCR Reaction

1. Prepare the reaction mixture:

Stock solution	Amount	Final Concentration
AMV/*Tfl* 5× reaction buffer	10 μl	1×
dNTP mix (10 m*M* of each dNTP)	1 μl	0.2 m*M* each
Forward primer	50 pmol	1 μ*M*
Reverse primer	50 pmol	1 μ*M*
25 m*M* MgSO$_4$	2 μl	1 m*M*
AMV reverse transcriptase (5U/μl)	1 μl	0.1 U/μl

2. Add one to two drops of nuclease-free mineral oil, or cap tubes if using a heated-lid thermal cycler, and place tubes in a thermal cycler. Run the following program:

First strand cDNA synthesis

1 cycle	48°C for 45 min
1 cycle	94°C for 2 min

Second strand cDNA synthesis and PCR

30 cycles	94°C for 30 min
	60°C for 1 min
	68°C for 2 min
1 cycle	68°C for 7 min
1 cycle	4°C until the sample is removed from the equipment

d. Analysis and Purification of the RT-PCR Reaction Product

Analyze the entire reaction product by 1% (w/v) agarose gel electrophoresis in 40 m*M* Tris-acetate, 1 m*M* EDTA (pH 8.0) (TAE) buffer using UV transillumination with ethidium bromide staining.

Purification of the RT-PCR cDNA product

1. Excise PCR product from the agarose gel under UV illumination.
2. Add three volumes of 6 *M* NaI (pH 7.0 to 7.4) to the gel block and incubate between 45 and 55°C for 5 min.
3. Add 3 μl of Glassmilk (Cat. No. 1001-400; BIO 101, La Jolla, CA) suspension and incubate for 5 min.
4. Pellet the Glassmilk/DNA complex for 5 s. Remove the supernatant and set aside.
5. Wash the pellet three times with wash buffer (0.1 *M* NaCl, 75% ethanol).
6. Elute DNA by adding 50 μl of nuclease-free water.

The sequence of the PCR product must always be confirmed by DNA sequencing. This can be done either by direct sequencing of the PCR product or after subcloning the cDNA into the pBluescript SK(-) vector (Stratagene).

2. Functional expression in Hep G2 cells

a. *Expression Vector Construction*

Preparation of the PLGFR cDNA fragment

1. Preparation of the PLGFR cDNA

PLGFR Insert:PCR DNA, 1 µg/µl	2 µl
Bgl II (10 U/µl)	1 µl
Not I (10 U/µl)	1 µl
Restriction enzyme buffer (10×)	2 µl
Nuclease-free water	14 µl

Incubate at 37°C for 2 h.

2. Extract the cDNA insert with phenol-chloroform (1:1; v/v) twice, followed by another chloroform extraction and ethanol precipitation. Resuspend the DNA in nuclease-free water.

3. Prepare the vector digestion reaction mixture:

Vector: pcDNA3.1 (1 µg/µl)	2 µl
Bam HI (10 U/µl)	1 µl
Not I (10 U/µl)	1 µl
Restriction enzyme buffer (10×)	2 µl
Nuclease-free water	14 µl

Incubate at 37°C for 2 h; extract the vector DNA as described for the insert above, and resuspend the vector DNA in nuclease-free water.

4. Ligate the plasmid vector and PLGFR cDNA insert.

Prepare the ligation mixture:

Vector DNA	100 ng
Insert DNA	50 ng
T4 DNA ligase (Weiss units)	1 U
Ligase buffer (10×)	1 µl
Nuclease-free water to final volume	10 µl

Incubate at 14°C for 16 h.

5. Transform bacteria with the recombinant pCDNA-PLGFR vector.

   a. Thaw a 200-µl aliquot of competent cells on ice.

   b. Add 1 µl of the ligation mixture to the competent cells, mix by gentle swirling, and incubate on ice for 30 min.

   c. Heat the tube to 42°C for 45 to 60 s; then cool on ice for 2 min.

   d. Add 0.8 ml of SOC medium and incubate for 1 h at 37°C with shaking at 150 rpm.

   e. Plate 100 to 200 µl of the transformation mixture on ampicillin selection plates.

Analysis of transformants by PCR

1. Randomly pick up ~10 bacterial transformants and transfer directly to PCR reaction tubes.

2. PCR reaction mixture:

Stock Solution	Amount	Final Concentration
Taq DNA polymerase reaction		
Buffer (10×)	5 µl	1×
dNTP Mix (10 mM of each dNTP)	1 µl	0.2 mM each
Forward primer	50 pmol	1 µM
Reverse primer	50 pmol	1 µM
25 mM MgCl$_2$	4 µl	2 mM
Taq DNA polymerase (5 U/µl)	1 µl	0.1 U/µl
Nuclease-free water		To a final volume of 50 µl

Overlay the reaction with nuclease-free mineral oil, or cap tubes for use in a heated-lid thermal cycler, and place tubes into the thermal cycler and run the following program:

1 cycle	95°C for 5 min
30 cycles	94°C for 45 s
	55°C for 45 s
	72°C for 1.5 min
1 cycle	72°C for 10 min

Analyze the entire reaction product by agarose gel electrophoresis as described above.

Preparation of plasmid cDNA

1. Inoculate 10 ml of Luria-Bertani (LB) medium containing ampicillin (100 µg/ml) with a single PCR-positive bacterial colony and incubate at 37°C with vigorous shaking for 12 to 16 h.

2. Transfer the culture to a 1.5-ml microcentrifuge tube and centrifuge at 12,000 × g for 1 min. Aspirate the supernatant, leaving the bacterial pellet as dry as possible. Resuspend the pellet by vortexing in 100 µl of ice-cold cell resuspension buffer; incubate for 5 min at room temperature

3. Add 200 µl of freshly prepared cell lysis buffer; mix by inversion and incubate for 5 min on ice.

4. Add 150 µl of ice-cold 3 M potassium acetate solution (pH 4.8); mix by inversion and incubate on ice for 5 min.

5. Centrifuge at 12,000 × g for 5 min; transfer the supernatant to a fresh tube and add 0.5 µl of 100 µg/µl DNase-free RNase A; incubate at room temperature for 5 min.

6. Extract the plasmid DNA with an equal volume of phenol:chloroform:isoamyl alcohol (25:24:1) twice.

7. Precipitate the upper aqueous phase with 2.5 × vol of chilled 100% ethanol on dry ice for 5 min. Centrifuge at 12,000 × g for 5 min.

8. Rinse the pellet with 70% ethanol, dry it under vacuum, and dissolve in nuclease-free water.

b. Transfection and Selection

Stable transfectants of Hep G2 cells, expressing the PSP24 and Edg receptors, can be established by electroporation with plasmid DNA carrying the receptor cDNA insert, using a BTX, model ECM 600, Electro Cell Manipulator.

Preparation of cells and electroporation

1. Harvest exponentially growing Hep G2 cells by detaching the cells with 0.05% trypsin/ 0.53 mM EDTA. Centrifuge and resuspend 1×10^7 cells in 300 μl of growth media (DMEM + 10% FBS).

2. Transfer the cells to a disposable electroporation cuvette (BTX, Part No. 620, 2 mm gap).

3. Add 10 μg of plasmid DNA (in 10 μl of water) to the cells and mix in the cuvette.

4. Set electroporator to T = 500 V, C = 1700 μF, R = R4, S = 100 V.

5. Place cuvette in the chamber and press A to activate.

6. Allow cells to rest at room temperature for 10 to 15 min.

Growth and selection

1. Dilute cells with 10 ml of growth medium and distribute among ten 100-mm dishes.

2. After 3 days, begin selection by adding 400 μg/ml of Geniticin (G418) to the normal growth medium.

3. Change medium as needed and continue selection with G418 until single colonies are visible (3 to 6 weeks).

4. When colonies are visible to the naked eye, pick them up with a sterile pipette tip and transfer to a 24-well plate.

5. Make a replica plate of the selected clones and screen by PCR.

c. Functional Analysis

Assaying the functional expression of the stable transfectants is done by monitoring the PLGF-mediated activation of the signal transduction pathways. Measuring changes in second messenger levels and/or the activation of downstream targets can help determine the specific coupling between PLGFRs and the different heterotrimeric G proteins. As described above, the activation of the G_q-mediated pathway can be monitored by measuring changes in intracellular calcium. The G_i-mediated pathways can be monitored by measuring changes in cAMP levels as well as in the activation of MAP kinases. The extracellular signal regulated kinase-1 (ERK1) and ERK2 MAP kinases are activated within 5 to 10 min following the application of PLGFs to cells, which can be measured by quantitating immune-complex kinase activity (Figure 8).

Monitoring PLGFR activity through ERK1 and ERK2 MAP kinases

1. Plate 5×10^5 cells in 60-mm dishes, in DMEM containing 10% FBS, and allow the cells to attach for 1 day.

Lipid Second Messengers

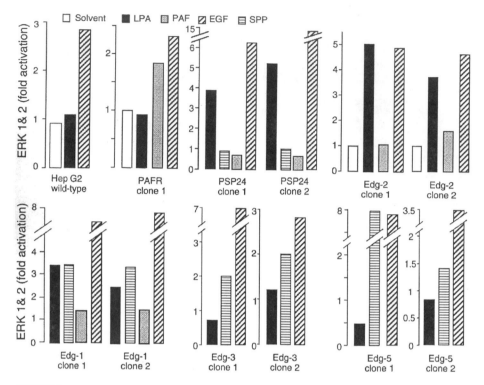

FIGURE 8

Heterologously expressed PLGFRs in Hep G2 cells introduce PLGF-mediated activation of ERKs. Hep G2 cells, stably transfected with the PSP24, Edg-1, Edg-2, Edg-3, Edg-5 and PAF receptors (PAFR) were exposed for 10 min to solvent control, 1 μM of LPA, sphingosine-1-phosphate (SPP), and PAF, or 50 ng/ ml of EGF. ERK1 and ERK2 activities were measured by immune-complex kinase assay as described in the text. PSP24 clones and Edg-2 clones show the activation of ERKs in response to LPA but not to PAF. In contrast, the PAFR clone 1 shows no change in ERK activity in response to LPA, but responds to its cognate ligand PAF. Edg-3 and Edg-5 expressing cells show ERK activation in response to SPP but not to LPA. Interestingly, Edg-1 expression in Hep G2 cells introduced responsiveness to both SPP and to a lesser extent to LPA.

2. Serum-starve the cells, in DMEM alone, for 24 h prior to treatment.

3. Treat the cells with a 1 μM concentration of the PLGFs for 10 min at 37°C.

4. Rinse the cells once with ice-cold PBS.

5. Add 500 μl of ice-cold lysis buffer, and rock at 4°C for 20 to 25 min.

6. Harvest cells and remove cell debris by centrifugation at $10,000 \times g$ for 10 min at 4°C.

7. Determine the protein concentration of the samples by protein assay (samples should be frozen at –80°C at this time if they are not to be used immediately).

8. Mix 200 μg of total cell protein with 2 μl of anti-ERK1, 2 μl of anti-ERK2 (Santa Cruz), and 20 μl of Protein G–Sepharose (Pharmacia Biotech, Piscataway, NJ) in a microcentrifuge tube and incubate for 2 h at 4°C with constant rotation. Adjust the sample volume to 200 μl with lysis buffer, depending on the volume of protein added.

9. Centrifuge and wash the immune complexes twice with 1 ml of lysis buffer and once with 1 ml of kinase buffer.

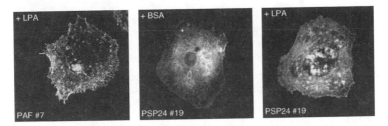

FIGURE 9
Heterologous expression of the PSP24 receptor in Hep G2 cells introduces LPA-induced stress fiber formation. Exposure of the cells expressing the PAFR (panel A) to 10 μM LPA for 15 min does not lead to actin polymerization as indicated by the lack of stress fibers. PSP24 clone 1 exposed to solvent (10 μM BSA, panel B) does not show stress fibers. In contrast, exposure of the same cells to LPA causes the formation of stress fibers traversing the cytoplasm (15 min, 10 μM). Rhodamine-phalloidin staining, imaging by confocal microscopy.

10. Drain the Protein G–Sepharose beads to complete dryness with a fine-gauge Hamilton syringe.

11. Begin the kinase reactions by adding 30 μl of kinase buffer containing 10 μM cold ATP, 2.5 μCi [γ^{32}P]ATP (10 Ci/mmol, Amersham), and 7 μg of myelin basic protein (Sigma).

12. Incubate the reactions for 30 min at 30°C and stop with the addition of 10 μl of 4 × SDS-PAGE sample buffer.

13. Resolve the phosphorylated proteins on a 14% SDS–polyacrylamide gel. Stain the gel with Coomassie Blue, dry, and quantitate either by cutting the bands from the gel or by phosphoimager analysis.

Functional expression of PLGFRs can be easily monitored for the formation of stress fibers in the stably transfected cells. This response is mediated through the heterotrimeric G protein subunits α$_{12/13}$ and the small GTPase RhoA (Figure 9).

Cell preparation, fixation, and permeabilization

1. Plate 5 × 10^4 cells onto 22 × 22 mm glass coverslips placed in a petri dish in growth medium and allow the cells to attach to the coverslips for at least 1 day.

2. Serum-starve the cells, in DMEM alone, for 16 to 24 h prior to treatment.

3. Treat the cells with a 1 to 10 μM concentration of the PLGFs for 15 to 30 min, at 37°C.

4. Remove the medium and immediately fix the cells in Dulbecco's PBS (DPBS) containing 3.7% formaldehyde for 10 min, at room temperature.

5. Rinse with DPBS.

6. Permeabilize the cells with 0.1% Triton X-100 in DPBS for 15 min at room temperature.

Staining and visualization of stress fibers

1. Rinse with DPBS containing 0.1% BSA.

2. Dilute rhodamine phalloidin (Molecular Probes, Cat. No. R-415) in DPBS containing 0.1% BSA to a final concentration of 5 U/ml.

3. Pipette 100 μl of rhodamine phalloidin onto the coverslip and incubate for 1 h. Keep coverslips in a wet chamber (petri dish with filter paper lining) and protect from light.

4. Rinse coverslips three times with DPBS containing 0.1% BSA, followed by two washes with DPBS.

5. Invert coverslips, mount on a microscope slide using one drop of Aqua-mount (Lerner Laboratories, Cat. No. 13600), and dry overnight in the dark.

6. Stress fibers are best observed by confocal fluorescence microscopy.

There is ample evidence that in addition to the PSP and Edg receptors, many more PLGFRs exist with distinct tissue distribution, signaling properties, and ligand specificities. The Hep G2 model may be a powerful technique in characterizing these novel receptors.

Reagents/Supplies Needed

Sources of Xenopus laevis

Adult female frogs can be purchased from

Carolina Biological Supply Company, 2700 York Road, Burlington, NC 27215, (800) 334-5551. Ordering information: Cat. No. P7-L1570.

NASCO, 901 Janesville Ave., Fort Atkinson, WI 53538, (920) 563-2446. Ordering information: Cat. No. LM00535M, females 9+ cm; or Cat. No. LM00531M, oocyte-positive females.

Xenopus I, 5654 Merkel Rd., Dexter, MI 48130, (313) 426-2083. Ordering information: Cat. No. 4216, oocyte-positive non-hormone-treated females; wild frogs are also available on request.

Primulin

Aldrich Chemical Co., Inc., Milwaukee, WI (Cat. No. 20,686)
Stock solution of 1%, w/v, primulin in deionized water. Dilute 100-fold in a mixture of acetone and deionized water (4:1, v/v).

Molybdenum Blue Spray Reagent

Sigma (Cat. No. M-3389)

Ca^{2+}-Free Ovarian Ringer's – 2 Solution (OR-2)

82.5 mM NaCl
2 mM KCl
1 mM MgCl$_2$
5 mM HEPES, pH 7.5, with NaOH

Defolliculation Solution

110 mM NaCl

1 mM EDTA
10 mM HEPES, pH 7.6, with NaOH

Modified Barth's Solution

88 mM NaCl
1 mM KCl
2.4 NaHCO$_3$
0.82 mM MgSO$_4$
0.33 mM Ca(NO$_3$)$_2$
0.41 mM CaCl$_2$
10 mM HEPES, pH 7.5 with NaOH
Sterilize by filtration; add 0.1 mg/ml gentamycin (Sigma G1272, 10 mg/ml stock).

Frog Ringer's Solution

120 mM NaCl
2 mM KCl
1.8 mM CaCl$_2$
5 mM HEPES, pH 7.0, with NaOH

Krebs Solution

120 mM NaCl
5 mM KCl
0.62 mM MgSO$_4$
1.8 mM CaCl$_2$
10 mM HEPES, pH 7.2
6 mM glucose

For *nominally Ca^{2+}-free Krebs*, replace CaCl$_2$ with

0.89 mM MgCl$_2$
1 mM EGTA

TAE Buffer

40 mM Tris-acetate, pH 8.0
1 mM EDTA

SOC Medium

20 g Bacto®-tryptone
5 g Bacto-yeast extract
0.5 g NaCl
20 mM glucose
In 1 l deionized water.

LB Medium

10 g Bacto-tryptone
5 g Bacto-yeast extract
5 g NaCl
Add deionized water to 1 l (pH 7.5).

Cell Lysis Buffer

0.2 N NaOH
1% SDS (w/v)

Immune Complex Lysis Buffer

20 mM Tris-HCl, pH 8.0
137 mM NaCl
1 mM EGTA
1% Triton X-100
10% glycerol
1.5 mM MgCl$_2$

Add just prior to use:

1 mM Na-vanadate
1 mM phenylmethylsulfonyl sulfide (PMSF)
10 μg/ml leupeptin
10 μg/ml aprotinin
50 mM NaF

Kinase Buffer

30 mM Tris-HCl, pH 8.0
20 mM MgCl$_2$
2 mM MnCl$_2$

III. Discussion

In this chapter, we described some of the elementary techniques used for the isolation and characterization of PLGFs and their receptors. This is a very rapidly developing field. It is obvious that there are many more endogenous PLGFs in different biologic fluids that have not yet been identified. It will be a future task to decipher the different enzymatic pathways that lead to the stimulus-coupled generation of these mediators. Moreover, there are a number of laboratories engaged in the development of synthetic PLGF agonists and antagonists with receptor subtype–specific action. It is likely that these attempts will identify many novel tools to probe PLGF function *in vitro* and *in vivo*. In addition, there is ample evidence that more PLGFRs exist and will be cloned in the future. For example, HEK 293,[2,25] Rat-1,[65,66] and Sp2-O-Ag14[54] cells do not express Edg-2, Edg-4, and PSP 24 receptors (Guo and Tigyi, unpublished), yet all of them respond to LPA. Consequently, it appears that the principles of PLGF signaling will be somewhat similar to those of the prostanoids and leukotrienes, which signal through a variety of receptors with overlapping selectivity.

References

1. **An, S., Dickens, M. A., Bleu, T., Hallmark, O. G., and Goetzl, E. J.,** Molecular cloning of the human Edg2 protein and its identification as a functional receptor for lysophosphatidic acid, *Biochem. Biophys. Res. Commun.*, 231, 619, 1997.

2. **Bittman, R., Swords, B., Liliom, K., and Tigyi, G.,** A short synthesis of lipid phosphoric acid analogues inhibiting the lysophosphatidate receptor, *J. Lipid Res.*, 37, 391, 1996.

3. **Buenemann, M., Brandts, B., Meyer zu Heringdorf, D., van Koppen, C. J., Jakobs, K. H., and Pott, L.,** Activation of muscarinic K^+ current in guinea pig atrial myocytes by sphingosine-1-phosphate, *J. Physiol.* (London), 489, 701, 1996.

4. **Buenemann, M., Liliom, K., Brandts, B., Pott, L., Tseng, J.-L., Desiderio, D. M., Sun, G., Miller, D., and Tigyi, G.,** A novel membrane receptor with high affinity for lysosphingomyelin and sphingosine 1-phosphate in atrial myocytes, *EMBO J*, 15, 5527, 1996.

5. **Christie, W. W.,** Detectors for high performance liquid chromatography of lipids with special reference to evaporative light-scattering detection, in *Advances in Lipid Methodology*, Christie, W. W., Ed., Vol. 1, The Oily Press, Ayr, 1992, 239.

6. **Chuprun, J. K., Raymond, J. R., and Blackshear, P. J.,** The heterotrimeric G protein $G_{\alpha i2}$ mediates lysophosphatidic acid-stimulated induction of the *c-fos* gene in mouse fibroblasts, *J. Biol. Chem.*, 272, 773, 1997.

7. **Coleman, A.,** Expression of exogenous DNA in *Xenopus oocytes*, in *Transcription and Translation—A Practical Approach*, Hames, B. D. and Higgins, S. J., Eds., IRL Press, Oxford, 1987, 49.

8. **Coleman, A.,** Transplantation of eukaryotic messenger RNA in *Xenopus oocytes*, in *Transcription and Translation—A Practical Approach*, Hames, B. D. and Higgins, S. J., Eds., IRL Press, Oxford, 1987, 271.

9. **Desai, N. N. and Spiegel, S.,** Sphingosylphosphorylcholine is a remarkably potent mitogen for a variety of cell lines, *Biochem. Biophys. Res. Commun.*, 181, 361, 1991.

10. **Dumont, J. N.,** Oogenesis in *Xenopus laevis* (Daudin), I. Stages of oocyte development in laboratory maintained animals, *J. Morphol.*, 136, 153, 1972.

11. **Durieux, M., Salafranca, M., Lynch, K., and Moorman, J.,** Lysophosphatidic acid induces a pertussis toxin-sensitive Ca^{2+}-activated Cl^- current in *Xenopus laevis* oocytes, *Am. J. Physiol. (Cell Physiol.)*, C263, C896, 1992.

12. **Dyer, D., Tigyi, G., and Miledi, R.,** The effect of serum albumin on PC12 cells. I. Neurite retraction and activation of the phosphoinositide second messenger system, *Mol. Brain Res.*, 14, 293, 1992.

13. **Eicholtz, T., Jalink, K., Fahrenfort, I., and Moolenaar, W. H.,** The bioactive phospholipid lysophosphatidic acid is released from activated platelets, *Biochem. J.*, 291, 677, 1993.

14. **Erickson, J. R., Wu, J. J., Goddard, G., Kawanishi, K., Tigyi, G., Tomei, L. D., and Kiefer, M. C.,** The putative lysophosphatidic acid receptor Edg-2/Vzg-1 functionally couples to the yeast response pathway, *J. Biol. Chem.*, 273, 1506, 1998.

15. **Ferguson, J. E. and Hanley, M. R.,** Phosphatidic acid and lysophosphatidic acid stimulate receptor-regulated membrane currents in the *Xenopus laevis* oocyte, *Arch. Biochem. Biophys.*, 297, 388, 1992.

16. **Fernhout, B., Dijcks, F., Moolenaar, W. H., and Ruigt, G.,** Lysophosphatidic acid induces inward currents in *Xenopus laevis* oocytes; evidence for an extracellular site of action, *Eur. J. Pharmacol.*, 213, 313, 1992.

17. **Fischer, D., Liliom, K., Guo, Z., Virag, T., Murakami-Murofushi, K., Kobayashi, S., Erickson, J. R., and Tigyi, G.,** Naturally occurring analogs of LPA elicit different cellular responses through selective activation of multiple receptor subtypes, *J. Biol. Chem., Mol. Pharmacol.* (in press).

18. **Fukami, K. and Takenawa, T.,** Phosphatidic acid that accumulates in platelet-derived growth factor-stimulated Balb/c 3T3 cells is a potential mitogenic signal, *J. Biol. Chem.*, 267, 10988, 1992.

19. **Gerrard, J. M. and Robinson, P.,** Identification of the molecular species of lysophosphatidic acid in platelets stimulated by thrombin, *Biochem. Biophys. Acta*, 1001, 282, 1989.

20. **Grynkiewicz, G., Poenie, M., and Tsien, R.,** A new generation of Ca^{2+} indicators with greatly improved fluorescence properties, *J. Biol. Chem.*, 260, 3440, 1985.

21. **Guo, Z., Liliom, K., Fischer, D. J., Bathurst, I. C., Tomei, D. J., Kiefer, M. C., and Tigyi, G.,** Molecular cloning of a high affinity receptor for the growth factor-like phospholipid mediator lysophosphatidic acid, *Proc. Natl. Acad. Sci. U.S.A.*, 93, 14367, 1996.

22. **Hecht, J. H., Weiner, J. A., Pots, S. R., and Chun, J.,** Vzg-1 encodes a lysophosphatidic acid receptor expressed in neurogenic regions of the developing cerebral cortex, *J. Cell. Biol.*, 135, 1071, 1996.

23. **Hooks, S. B., Ragan, S. P., Hopper, D. W., Honeman, C. W., Durieux, M. E., MacDonald, T. L., and Lynch, K. R.,** Characterization of a receptor subtype-selective lysophosphatidic acid mimetic, *Mol. Pharmacol.*, 53, 188, 1998.

24. **Ilyin, V. and Parker, I.,** Effects of alcohols on responses evoked by inositol trisphosphate in *Xenopus* oocytes, *J. Physiol.* (London), 448, 339, 1992.

25. **Imamura, F., Horai, T., Mukai, M., Shinkai, K., Sawada, M., and Akedo, H.,** Induction of *in vitro* tumor cell invasion of cellular monolayers by lysophosphatidic acid or phospholipase D, *Biochem. Biophys. Res. Commun.*, 193, 497, 1993.

26. **Jalink, K., Eicholtz, T., Postma, F. R., v. Corven, E. J., and Moolenaar, W. H.,** Lysophosphatidic acid induces neuronal shape changes via a novel, receptor-mediated signaling pathway: similarity to thrombin action, *Cell Growth Diff.*, 4, 206, 1993.

27. **Jalink, K., Moolenaar, W. H., and Van Duijn, B.,** Lysophosphatidic acid is a chemoattractant for *Dictyostelium discoideum* amoebae, *Proc. Natl. Acad. Sci. U.S.A.*, 90, 1857, 1993.

28. **Lapetina, E., Billah, M., and Cuatrecasas, P.,** Lysophosphatidic acid potentiates the thrombin-induced production of arachidonate metabolites in platelets, *J. Biol. Chem.*, 256, 11984, 1981.

29. **Lapetina, E. and Cuatreacasas, P.,** Stimulation of phosphatidic acid production in platelets precedes the formation of arachidonate and parallels the release of serotonin, *Biochem. Biophys. Acta,* 573, 394, 1979.

30. **Lattanzio, F. A. and Bartschat, D. K.,** The effect of pH on rate constants, ion selectivity and thermodynamic properties of fluorescent calcium and magnesium indicators, *Biochem. Biophys. Res. Commun.,* 177, 184, 1991.

31. **Liliom, K., Bittman, R., Swords, B., and Tigyi, G.,** *N*-palmitoyl-serine and *N*-palmitoyl-phosphoric acids are selective competitive antagonists of the lysophosphatidic acid receptors, *Mol. Pharmacol.,* 50, 616, 1996.

32. **Liliom, K., Fischer, D. J., Sun, G., Miller, D. D., Tseng, J.-L., Desiderio, D. M., Seidel, M. C., Erickson, J. R., and Tigyi, G.,** Identification of a novel growth factor-like lipid: 1-*O*-*cis*-alk-1′-enyl-2-lyso-*sn*-glycero-3-phosphate (Alkenyl-GP) that is present in commercial sphingolipid preparations, *J. Biol. Chem.,* 273, 13461, 1998.

33. **Liliom, K., Guan, Z., Tseng, J. L., Desiderio, D. M., Tigyi, G., and Watsky, M.,** Growth factor-like phospholipids generated following corneal injury, *Am. J. Physiol. (Cell Physiol.),* 43, 274, CXXX, C1065, 1998.

34. **Liliom, K., Murakami-Murofushi, K., Kobayashi, H., Murofushi, H., and Tigyi, G.,** *Xenopus* oocytes express multiple receptors for LPA-like lipid mediators, *Am. J. Physiol. (Cell Physiol.),* 39, 772, 1996.

35. **Miledi, R., Parker, I., and Sumikawa, K.,** Transplanting receptors from brains into oocytes, *Fidia Res. Found. Neurosci. Award Lect.,* 3, 57, 1989.

36. **Miledi, R. and Woodard, R. M.,** Membrane currents elicited by prostaglandins, atrial natriuretic factor and oxytocin in follicle-enclosed *Xenopus* oocytes, *J. Physiol. (London),* 416, 623, 1989.

37. **Moolenaar, W. H., Kranenburg, O., Postma, F. R., and Zondag, G.,** Lysophosphatidic acid: G protein signaling and cellular responses, *Curr. Opin. Cell Biol.,* 9, 168, 1997.

38. **Murakami-Murofushi, K., Kaji, K., Kano, K., Fukuda, M., Shioda, M., and Murofushi, H.,** Inhibition of cell proliferation by a unique lysophosphatidic acid, PHYLPA, isolated from *Physarum polycephalum*: signaling events of antiproliferative action by PHYLPA, *Cell Struct. Funct.,* 18, 363, 1993.

39. **Nomura, H. and Ueda, H.,** Reconstitution of lysophosphatidic acid receptor and G proteins using the baculoviral expression system, *Abstr. Biochem. Soc., Ann. Mtg. Jpn.,* 69, 696, 1997.

40. **Parker, I. and Miledi, R.,** Changes in intracellular calcium and in membrane currents evoked by injection of inositol triphosphate into *Xenopus* oocytes, *Proc. R. Soc. London B,* 228, 307, 1986.

41. **Parker, I., Sumikawa, K., and Miledi, R.,** Activation of a common effector system by different brain neurotransmitter receptors in *Xenopus* oocytes, *Proc. R. Soc. London,* 231, 37, 1987.

42. **Quehenberger, O., Prossnitz, E. R., Cochrane, C. G., and Ye, R. D.,** Absence of G(i) proteins in the SF9 insect cell. Characterization of the uncoupled recombinant N-formyl peptide receptor, *J. Biol. Chem.,* 267, 19757, 1992.

43. **Radin, N. S.,** Lipid extraction, in *Neuromethods,* Vol. 7, Boulton, A. A., Baker, G. B., and Horrocks, L. A., Eds., Humana, Clifton, NJ, 1988, 1.

44. **Ridley, A. J. and Hall, A.,** The small GTP-binding protein rho regulates the assembly of focal adhesions and actin stress fibers in response to growth factors, *Cell*, 70, 389, 1992.

45. **Shiono, S., Kawamoto, K., Yoshida, N., Kondo, T., and Inagami, T.,** Neurotransmitter release from lysophosphatidic acid stimulated PC12 cells: involvement of lysophosphatidic acid receptors, *Biochem. Biophys. Res. Commun.*, 193, 667, 1993.

46. **Soreq, H. and Seidman, S.,** *Xenopus* oocyte microinjection: from gene to protein, in *Meth. Enzymol.*, Vol. 207, Rudy, B. and Iverson, L. E., Eds., Academic Press, San Diego, 1992, 225.

47. **Stuhmer, W.,** Electrophysiological recording from *Xenopus* oocytes, in *Ion Channels*, Vol. 207, Rudy, B. and Iverson, L. E., Eds., Academic Press, San Diego, 1992, 353.

48. **Sumikawa, K., Parker, I., and Miledi, R.,** Expression of neurotransmitter receptors and voltage-activated channels from brain mRNA in *Xenopus* oocytes, in *Methods in Neuroscience*, Vol. 1, *Genetic Probes*, Conn, P. M., Ed., Academic Press, Miami, 1989, 30.

49. **Tigyi, G., Dyer, D., Matute, C., and Miledi, R.,** A serum factor that activates the phosphatidylinositol phosphate signaling system in *Xenopus* oocytes, *Proc. Natl. Acad. Sci. U.S.A.*, 87, 1521, 1990.

50. **Tigyi, G., Dyer, D., and Miledi, R.,** Lysophosphatidic acid possesses dual action in cell proliferation, *Proc. Natl. Acad. Sci. U.S.A.*, 91, 1908, 1994.

51. **Tigyi, G., Fischer, D., Sebök, Á., Marshall, F., Dyer, D., and Miledi, R.,** Lysophosphatidic acid-induced neurite retraction in PC12 cells: neurite protective effects of cyclic AMP signaling, *J. Neurochem.*, 66, 549, 1996.

52. **Tigyi, G., Fischer, D., Yang, C., Dyer, D., Sebök, Á., and Miledi, R.,** Lysophosphatidic acid-induced neurite retraction in PC12 cells: control by phosphoinositide-Ca^{2+} signaling and Rho, *J. Neurochem.*, 66, 537, 1996.

53. **Tigyi, G., Henschen, A., and Miledi, R.,** A factor that activates oscillatory chloride currents in *Xenopus* oocytes copurifies with a subfraction of serum albumin, *J. Biol. Chem.*, 266, 20602, 1991.

54. **Tigyi, G., Hong, L., Shibata, M., Parfenova, H., Yakubu, M., and Leffler, C.,** Lysophosphatidic acid alters cerebrovascular reactivity in piglets, *Am. J. Physiol. (Heart Circ. Physiol.)*, 37, H2048, 1995.

55. **Tigyi, G. and Miledi, R.,** Lysophosphatidates bound to serum albumin activate membrane currents in *Xenopus* oocytes and neurite retraction in PC12 pheochromocytoma cells, *J. Biol. Chem.*, 267, 21360, 1992.

56. **Tigyi, J., Tigyi, G., Liliom, K., and Miledi, R.,** Local anaesthetics inhibit receptors coupled to phosphoinositide signaling in *Xenopus* oocytes, *Pfluegers Arch. Eur. J. Physiol.*, 433, 478, 1997.

57. **Tokumura, A.,** A family of phospholipid autacoids: occurrence, metabolism, and bioactions, *Prog. Lipid Res.*, 34, 151, 1995.

58. **Tokumura, A., Fukuzawa, K., Yamada, S., and Tsukatani, H.,** Stimulatory effect of lysophosphatidic acids on uterine smooth muscles of non-pregnant rats, *Arch. Int. Pharmacodyn.*, 245, 74, 1980.

59. **Tokumura, A., Iimori, M., Nishioka, Y., Kitahara, M., Sakashita, M., and Tanaka, S.,** Lysophosphatidic acids induce proliferation of cultured vascular smooth muscle cells from rat aorta, *Am. J. Physiol. (Heart Circ. Physiol.)*, 267, C204, 1994.

60. Umansky, S. R., Shapiro, J. P., Cuenco, G. M., Foehr, W. M., Bathurst, I. C., and Tomei, L. D., Prevention of rat neonatal cardiomyocyte apoptosis induced by stimulated *in vitro* ischemia and reperfusion, *Cell Death Diff.*, 4, 608, 1997.

61. van Corven, E. J., Groenink, A., Jalink, K., Eichholtz, T., and Moolenar, W. H., Lysophosphatidicacid-induced cell proliferation: identification and dissection of signaling pathways mediated by G proteins, *Cell*, 59, 45, 1989.

62. van Corven, E. J., van Rijswijk, A., Jalink, K., van Der Bend, R. L., van Blitterswijk, W. J., and Moolenaar, W. H., Mitogetic action of lysophosphatidic acid on fibroblasts, *Biochem. J.*, 281, 163, 1992.

63. Wu, Y., Tomei, L. D., Bathurst, I. C., Zhang, F., Hong, C. B., Issel, C. J., Columbano, A., Salley, R. K., and Chien, S., Antiapoptotic compound to enhance hypothermic liver preservation, *Transplantation*, 63, 803, 1997.

64. Xu, Y., Casey, G., and Mills, G. B., Effect of lysophospholipids on signaling in the human Jurkat T cell line, *J. Cell. Physiol.*, 163, 441, 1995.

65. Xu, Y., Fang, X. J., Casey, G., and Mills, G. B., Lysophospholipids activate ovarian and breast cancer cells, *Biochem. J.*, 309, 933, 1995.

66. Yakubu, M. A., Liliom, K., Tigyi, G., and Leffler, C. W., Role of lysophosphatidic acid in endothelin-1- and hematoma-induced alteration of cerebral microcirculation, *Am. J. Physiol. (Reg. Comp. Integ. Physiol.)*, 42, R703, 1997.

67. Zhou, D., Luini, W., Bernasconi, S., Diomede, L., Salmona, M., Mantovani, A., and Sozzani, S., Phosphatidic acid and lysophosphatidic acid induce haptotactic migration of human monocytes, *J. Biol. Chem.*, 270(43), 25549, 1995.

68. Zondag, G. C. M., Postma, F. R., Verlaan, E. I., and Moolenaar, W. H., Sphingosine-1-phosphate signaling through G protein-coupled receptor edg-1, *Biochem. J.*, 330, 605, 1998.

Chapter 5

Leukotriene and
Lipoxin Biosynthesis

*Bruce D. Levy, Karsten Gronert, Clary B. Clish,
and Charles N. Serhan*

Contents

83

I.   Introduction

Leukotrienes (LT) and lipoxins (LX) are critical effectors in the control of diverse physiological and pathophysiological events, such as inflammation, reperfusion injury, and tissue repair.[1] LTs stimulate leukocyte extravasation and activation (LTB_4) and are potent vaso- and bronchoconstrictors (LTC_4 and LTD_4).[2] In contradistinction to these proinflammatory LT-mediated signals, LXs display leukocyte-selective actions that are characteristic of "stop signals" for acute inflammation. Namely, LXs inhibit LT-mediated activation of human neutrophils and stimulate human monocyte locomotion. Monocytes, but not neutrophils, have the capacity to metabolize LX to biologically less active products, thereby leading to their inactivation (reviewed in Reference 3). The timing and balance of LT and LX biosynthesis likely influences the intensity and duration of cellular responses during inflammation and other complex host responses.

Many different circulating and resident cell types likely participate in the biosynthesis of LT and LX. These lipoxygenase (LO)-derived eicosanoids are products of the enzymatic oxygenation of arachidonic acid (Figure 1). Arachidonate is esterified in the *sn*-2 position of membrane phospholipids and released upon cellular activation. Free arachidonic acid can then be converted by the action of 5-lipoxygenase to an unstable epoxide, LTA_4. Both arachidonic acid and LTA_4 (roughly 50%) can be released from its cell of origin to become available as a substrate for further conversion by neighboring cells to multiple bioactive products including the LTs (e.g., LTB_4, LTC_4) and LXs (e.g., LXA_4 and LXB_4).[4] LT-forming enzymes, such as LTA_4 hydrolase and LTC_4 synthase, are present in certain cell types without 5-lipoxygenase activity, and therefore donation of biosynthetic intermediates (e.g., LTA_4) from other cellular sources is required for LT generation by these cells.

Because the three mammalian LOs (5-, 12-, and 15-LO) are generally segregated into different cell types, LX biosynthesis also occurs via the bidirectional transfer of eicosanoid intermediates during cell–cell interactions (see Figure 1). There are three well-described pathways for LX transcellular biosynthesis that can be operative either independently or in concert during cell–cell interactions as cell types with different LO activity are assembled during multicellular host responses: the 5-LO initiated pathway, the 15-LO initiated pathway, and the aspirin-triggered cyclooxygenase-2 (COX-2) pathway (Figure 2, reviewed in Reference 3). Transcellular pathways for LT and LX biosynthesis amplify the levels of these LO-derived eicosanoids within a local milieu and result in the generation of novel bioactive products that are beyond the biosynthetic capacity of the individual cell types in isolation (Figure 3).

Powerful tools for further analysis of the roles of LT and LX in physiological and pathophysiological processes are rapidly evolving. Biosynthetic enzymes for LT

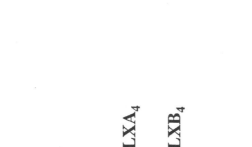
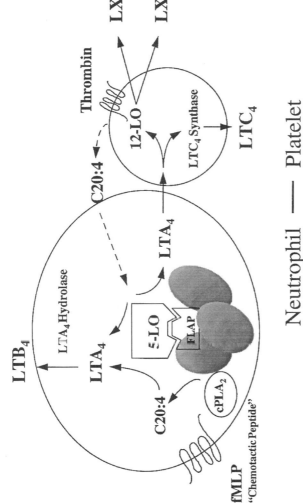

FIGURE 1

Transcellular biosynthesis of leukotrienes and lipoxins. Platelet–neutrophil interactions as a model of transcellular biosynthesis of leukotrienes and lipoxins. Cells activated by receptor mediated stimuli (e.g., fMLP and thrombin) release arachidonic acid (C20:4) from cell membranes for further metabolism within or outside the cell of origin of C20:4. The 5-LO product, LTA_4, is a pivotal intermediate with many potential fates, including enzymatic conversion to LTB_4 by neutrophil LTA_4 hydrolase or LTC_4 or LXs by platelet LTC_4 synthase or 12-LO, respectively. Note the potential for bidirectional exchange of intermediates in the donation of platelet C20:4 to neutrophil 5-LO for conversion to LTA_4, which is then exported back to the platelet for LT or LX formation.

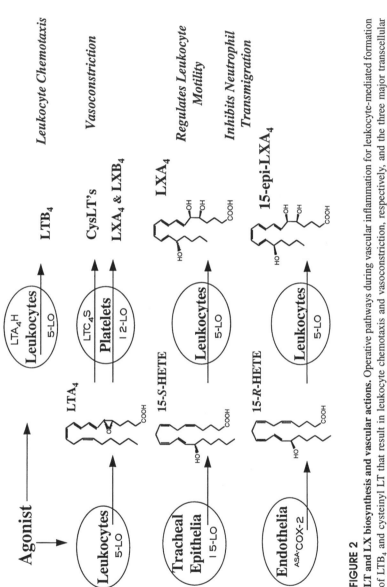

FIGURE 2

LT and LX biosynthesis and vascular actions. Operative pathways during vascular inflammation for leukocyte-mediated formation of LTB_4 and cysteinyl LT that result in leukocyte chemotaxis and vasoconstriction, respectively, and the three major transcellular routes for LX production, which carry anti-inflammatory properties by downregulating leukocyte motility. LO = lipoxygenase, COX = cyclooxygenase, $LTA_4H = LTA_4$ hydrolase, and $LTC_4S = LTC_4$ synthase.

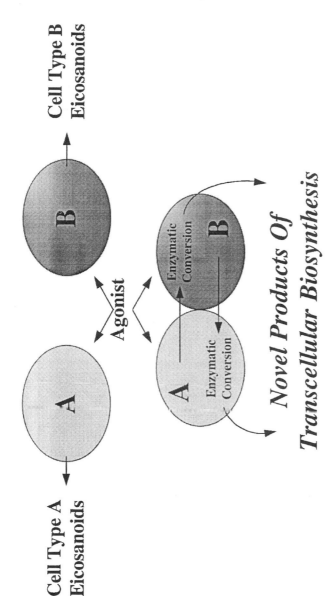

FIGURE 3

General considerations for investigating transcellular biosynthesis. Thorough investigation of transcellular eicosanoid biosynthesis requires an organized experimental approach. Consider the individual cell type (A or B) resting and activated profile of eicosanoid production alone prior to coincubation. During coincubation (A and B), stimulate with cell type–specific (A or B) agonists in parallel with incubations using costimulation (A and B) to determine the contribution of the individual cell types to the products of transcellular biosynthesis (e.g., leukotrienes and lipoxins). All of this information is required to elucidate the operative transcellular eicosanoid biosynthetic pathways.

and LX, such as 5-, 12-, and 15-LO and LTC$_4$ synthase, have been cloned.[5,6] The first LO-derived eicosanoid receptor to be cloned, the LXA$_4$ receptor (LXAR), was identified as a member of the family of G protein–linked receptors with seven transmembrane-spanning domains. Recently, a similar cloning strategy resulted in the identification of an LTB$_4$ receptor, another G protein–linked receptor closely related to the LXAR.[7] Currently, investigators are pursuing the gene encoding the LTD$_4$ receptor, which has already been pharmacologically characterized.[8] Identification of these receptors is crucial to elucidating the signal transduction pathway of their respective ligands and facilitates rational drug design.

In addition, mice without 5-LO or 12-LO activity and rabbits overexpressing 15-LO have been engineered to allow *in vivo* investigation of the roles of these important enzymes. Of interest, 5-LO-deficient mice show diminished airway inflammation and bronchial hyperreactivity after exposure to ovalbumin,[9] while rabbits overexpressing 15-LO were protected from the generation of atherosclerotic plaque and vascular inflammation despite being fed a diet rich in fat and cholesterol.[10] These studies, which demonstrate a diminished inflammatory response in mice without the capacity to generate LT and confirm the postulated anti-inflammatory role for 15-LO-derived eicosanoids, such as LX exemplify the breadth of investigation possible in this field of research.

This chapter reviews the diverse methodology required to determine the biosynthesis of LT and LX during cell–cell interactions, including current techniques for (1) cell isolation and incubation, (2) eicosanoid extraction, (3) detection of lipoxygenase expression, and (4) quantitation of LT and LX by both physical and immunologic means. Essential steps for each required technique are described along with hints for those new to this area of research. Composition of the needed buffers and sources for the reagents are listed under *Reagents Needed.*

II.   Protocol

A.   Cell Isolation

Diverse circulating and tissue-resident cell types have been demonstrated to participate in transcellular eicosanoid biosynthesis. Review of all possible combinations of cell types that are likely to occur *in vivo* is beyond the scope of this chapter; however, platelet–neutrophil interactions, known to occur *in vivo* at sites of vascular inflammation or injury and whose LT and LX transcellular biosynthetic pathways are well characterized, will be presented here to exemplify important issues in experimental design.

1.   Preparation of platelet-rich plasma from whole blood

1.   For each experiment, obtain fresh peripheral blood (usually ~120 to 180 ml) in 10% acid citrate dextrose (ACD; when autologous platelets and neutrophils are isolated) by venipuncture from healthy volunteer donors who have denied taking medications (prescription or other) for at least 1 week.

2.   Centrifuge (200 g, 25°C, 15 min) to obtain platelet-rich plasma.

2.   Platelet isolation

1.   Transfer platelet-rich plasma using a polypropylene transfer pipette into a 50-ml polypropylene tube.

Note: *Do not disturb the interface.*

2.   Dilute with HEPES-Tyrode buffer plus EDTA (final concentration 13.6 mM to prevent platelet activation) to 50 ml.

3.   The platelets are washed three times in HEPES-Tyrode buffer plus EDTA by centrifugation (800 g, 15 min, 25°C). Resuspend platelets in approximately 1 ml using gentle puffing with a polypropylene transfer pipette.

Note: *Take care to not dislodge the cell pellet or resuspend the ring of red erythrocytes evident in the pellet.*

Transfer platelets to a clean 50-ml polypropylene test tube and dilute in 25 ml HEPES-Tyrode buffer prior to repeat centrifugation.

4.   After the last wash, resuspend the platelets in phosphate-buffered saline (PBS) (without calcium or magnesium), pH 7.40 containing 13.6 mM EDTA and enumerate using a Coulter counter at a 1:20 dilution (cells:PBS + EDTA, v/v).

3.   Neutrophil isolation

1.   Neutrophils are isolated from whole blood after removal of the platelet-rich plasma (see step 1.2). Dilute with PBS plus 6% dextran at a ratio of 3:2:1, v/v/v, blood:PBS:dextran. Erythrocytes are allowed to sediment (25°C, 20 to 30 min) in a polypropylene graduated cylinder.

Note: *Cover with parafilm to avoid contamination.*

2.   Remove the upper layer (be careful not to remove excess erythrocytes at the interface) into 2 × 50 ml sterile polypropylene test tubes. Slowly underlay 10 ml of lymphocyte separation medium (LSM; Ficoll/Diatrizoate gradient) (Organon Teknika Corp., Durham, NC) to the bottom of each tube below the cell suspension.

Note: *Do not mix. There should be an undisturbed interface between the LSM and leukocyte suspension.*

Centrifuge (500 g, 30 min, 25°C) without brake.

3.   Collect the buffy coat enriched with mononuclear cells (upper opaque band) if desired. Otherwise, neutrophils are found in the cell pellet. Aspirate supernatant carefully.

4.   To remove the remaining erythrocytes, perform hypotonic lysis by resuspending the cell pellets in 45 ml distilled and deionized water for 30 s.

Note: *Count exactly because ≥30 s can result in leukocyte lysis!*

Follow by 5 ml of 10× HBSS. Mix gently.

5. Wash neutrophils twice in PBS by centrifugation (500 *g*, 25°C, 10 min) to remove RBC ghosts.

6. Resuspend in buffer of choice (usually 5 to 10 ml) and count using a hemocytometer.

B. Cell Incubation Conditions

The capacity of individual cell types to generate LT and/or LX can be evaluated, and, if cell–cell interactions are likely to occur during physiological or pathophysiological states, conditions can be simulated *ex vivo* to determine the presence of transcellular pathways for eicosanoid formation (see Figure 2). When investigating LT and LX transcellular biosynthesis during cell–cell interactions, it is essential to examine eicosanoid profiles during each of the following incubation conditions:

- Basal profile of individual cell type (i.e., without agonist);
- Individual cell type after activation by cell type–specific agonist;
- Stimulation of cells during coincubation using individual cell type–specific agonist;
- Costimulation of both cell types during coincubation using shared
    a. Receptor-independent stimuli,
    b. Receptor-mediated stimuli.

Healthy individuals typically have 1.8 to 7.7 × 10^9 neutrophils (54% total leukocytes)/L and 130 to 400 × 10^9 platelets/L of whole blood in the peripheral venous circulation. The ratio of neutrophils to platelets markedly increases at sites of vascular inflammation, as these cell types physically interact and total neutrophil numbers dramatically rise. In the presence of local agonists, such as thrombin and chemotactic peptides, and circulating agents, such as granulocyte/monocyte colony–stimulating factor (GM-CSF) and other cytokines, platelet–neutrophil interactions lead to the elaboration of lipid mediators that modulate the inflammatory host response and mediate the repair of tissue injury. Dissecting the products, their biosynthesis, and the factors that regulate their formation requires an organized experimental approach.

1. Sources of arachidonate

To determine the enzymatic potential of distinct cell types, incubations with exogenous arachidonate or biosynthetic intermediates can be carried out. The large amount of exogenous substrate (usually >10 μ*M*) may facilitate product detection, but often results in cell activation by a detergent-like effect and will subsume any attempts to regulate cell function. With available technology, such as reverse-phase high-pressure liquid chromatography (RP-HPLC) and liquid chromatography tandem mass spectrometry (LC/MS/MS), that enables detection of LT and LX in the picogram range (see Section II.D.2.b), eicosanoid biosynthesis without added

TABLE 1
Stimuli for Leukotriene and Lipoxin Formation

Agent	Activation of Pathway	Increased Expression
I. Phagocytic Stimuli		
Zymosan	5-LO	
IgE	5-LO	
Aggregated Ig	5-LO	
Monosodium urate crystals	5-LO	
II. Soluble Stimuli		
fMLP	5-LO	
C5a	5-LO	
Plasmin	5-LO	
Platelet-activating factor	5-LO	
TNFα	5-LO	
NGF	5-LO	
Vitamin D		5-LO and LTC_4 synthase
GM-CSF		5-LO, FLAP, and LTC_4 synthase
Interleukin-3		5-LO, FLAP, and LTC_4 synthase
TGF-β		5-LO and FLAP
Glucocorticoids		5-LO and FLAP
LPS		5-LO and FLAP
Oxidized LDL		FLAP
Thrombin	Platelet 12-LO	
EGF		Platelet 12-LO
Angiotensin II		Leukocyte 12-LO
Interleukin-4 and -13	15-LO	15-LO

Abbreviations: Ig = immunoglobulin, fMLP = formyl-methionyl-leucyl-phenylalanine, TNF = tumor necrosis factor, NGF = nerve growth factor, GM-CSF = granulocyte/macrophage colony–stimulating factor, LPS = lipopolysaccharide, LDL = low-density lipoprotein, LO = lipoxygenase, FLAP = five lipoxygenase–activating protein.

arachidonate (more closely resembling the *in vivo* state) should be analyzed so that product formation can be determined from available endogenous stores of arachidonate.

2.   Stimuli

As diverse as the cell types available for study are, so too is the array of stimuli that investigators have examined for LT and LX formation. The range of agonists that have been studied includes phagocytic stimuli (e.g., opsonized zymosan, immune complexes, and urate crystals) and soluble stimuli (e.g., chemotactic peptides, thrombin, calcium ionophores, complement components, and platelet-activating factor and cytokines) (Table 1). Notably, the biosynthetic profile for LT and LX formation during coincubation of differing cell types is unique to the agonists used to activate the cells. The most physiologically relevant stimuli are those that mediate their effects via a cellular receptor (e.g., thrombin or GM-CSF), while stimuli that bypass membrane receptors (e.g., calcium ionophores) are potent cellular agonists yet often result in a profile of eicosanoid products that is markedly different, qualitatively and quantitatively, from receptor-mediated stimuli.

To determine the eicosanoid profile of an individual cell type, it is often useful to expose the cell, in isolation or during coincubation, to cell type–specific agonists. For example, expose a neutrophil (PMN)-platelet (PLT) incubation to thrombin to determine the enzymatic contribution of the platelet. Incremental information can next be derived coincubations of cells in the presence of a cell type–specific agonist for PMN, such as the chemotactic peptide formyl-methionyl-leucyl-phenylalanine (fMLP). Last, coincubation of the two cell types with both relevant agonists often yields both increased amounts of the cell type–specific products and generation of *novel* products beyond the enzymatic capacity of either cell type in isolation (Figure 3).

Consider labeling each cell type independently with radioactive substrate to determine (1) if arachidonate or a biosynthetic intermediate is transferred between cells and (2) which enzymes are operative in each cell type in the generation of bioactive products (see Section II.D.2.a).

3.   Inhibitors

Multiple compounds are commercially available to inhibit cyclooxygenases 1 and 2 and most mammalian lipoxygenases. Many of these compounds were designed for the selective inhibition of an individual enzyme; however, most will nonspecifically affect related enzymes, especially at high concentrations. For this reason, incubation in the presence of inhibitors often results in supportive, but not conclusive results. In addition, although a compound may inhibit an enzyme from converting its substrate to the predominant product, its action may result in the formation of alternative products (e.g., aspirin inhibits the formation of endoperoxide by COX-2, yet triggers the generation of 15-R-hydroxyeicosatetraenoic acid (15-R-HETE) from arachidonate). Examples of COX and LO inhibitors are given in Table 2.

4.   Terminating the incubations

Cellular incubations may be terminated by either the use of an organic solvent or hyperosmolar solution to determine the total profile of LT and LX generated. To determine released vs. intracellular products, coincubations can be stopped by immersion into a 0°C water bath and centrifugation (800 g, 5 min, 0°C), which separates the cell pellet from the supernatant.

5.   Example: Platelet–Neutrophil coincubation

Given these general considerations, an example of a PMN-PLT coincubation might proceed as follows:

TABLE 2
Commonly Used Inhibitors of Leukotriene and Lipoxin Biosynthesis[a]

Agent	Common Range of Concentrations, μM	Enzyme Target
Aspirin	100–500	COX-1 and COX-2[b]
Indomethacin	100	COX-1 and COX-2
NS398	50–100	COX-2
17-Octadecynoic acid (ODYA)	500	Cytochrome p450
Nordihydroguaiaretic acid (NDGA)	0.2–50	5-,12-, and 15-LO
Caffeic acid	5-10	5-LO
Esculetin	100	12-LO
AACOCF3	15	Cytosolic phospholipase A_2 ($cPLA_2$)
5,8,11,14-Eicosatetraynoic acid (ETYA)	10–50	COX, 5-,12-, 15-LO, and $cPLA_2$

[a] These agents are representative of the wide array of inhibitors that are commercially available and can be utilized for the preferential inhibition of selected LT and LX biosynthetic enzymes.

[b] While aspirin inhibits COX-mediated prostaglandin formation, it also triggers the generation of 15-*R*-HETE by the acetylated enzyme. This feature is unique to aspirin and not a property of COX inhibition by nonsteroidal anti-inflammatory drugs in general.

Procedure

1. Incubate freshly isolated PMN—10 to 25×10^6 cells/ml PBS plus $CaCl_2$ (0.6 mM) and $MgCl_2$ (1 mM)—pH 7.4 in the presence or absence of recombinant human granulocyte/macrophage colony–stimulating factor (GM-CSFrh) (200 pM) for 90 min at 37°C.

2. Combine aliquots of neutrophils (PMN) (0.5 ml) with aliquots of platelets (PLT) (0.5 ml) and adjust the cell ratio from 0:1 to 100:1 (PLT:PMN) in parallel incubations to determine the contribution of PLT enzymes to LT and LX formation during coincubation.

3. Warm PMN and PLT for 5 min at 37°C before exposure to thrombin (1 U/ml), fMLP (10^{-7} M) or both soluble stimuli for 20 min at 37°C.

4. Stop incubations by addition of 1 volume of 0.05 M Tris-HCl (pH 8.0) and vigorous mixing.

5. After termination of the reaction, materials should be kept at –20°C for at least 60 min prior to analysis to facilitate protein precipitation, as well as to slow potential autoxidation.

6. Samples will remain stable for longer periods at –80°C, but should be extracted and analyzed within 24 to 48 h.

C. Eicosanoid Extraction

To determine LT and LX transcellular biosynthesis, the eicosanoids must first be extracted from the reaction materials before they can be concentrated and analyzed.

Eicosanoid extraction is complicated by obligate loss of compound. To control for the loss of unknown total amounts of LT and LX, a closely related compound that would *not* be an anticipated product of the coincubation should be used as an internal standard. It should be added to the reaction materials at the time the coincubation is terminated. Two major extraction protocols have been optimized for the recovery of LT and LX: liquid–liquid extraction and solid-phase extraction.

1.   Liquid–liquid extraction

1. Terminate coincubation with ≥ 2 volumes of methanol. After protein precipitation at $-20°C$ for at least 60 min, centrifuge (800 g, 15 min, 0°C).

2. Transfer supernatant (S_{800}) to a separatory funnel. Resuspend pellet in ≥ 2 volumes ethanol and repeat centrifugation. Pool S_{800} into the separatory funnel. Repeat wash with MeOH and pool S_{800}.

3. Add an equal volume of double-deionized water (ddH_2O) to the pooled S_{800}. Next add a large amount of ether (~200 ml).

Note: Be certain to work in an approved chemical fume hood.

4. Acidify the lower (i.e., aqueous) phase to pH 1 to 2. Add 0.1 N HCl drop-by-drop and check the pH of the lower phase by pH paper after vigorous mixing.

5. After pH adjustment, shake and ventilate separatory funnel for 1 min. Remove cap and allow interface to form.

6. Remove lower phase (ddH_2O and ethanol layer) into a second separatory funnel. Add another ~200 ml ether. Check and adjust pH. Save first funnel containing ether.

7. Shake second separatory funnel for 1 min. Remove and discard lower phase. Combine ether phases.

8. Wash ether phases with ~10 ml ddH_2O, shake well, and remove lower phase after interface has formed. Repeat until the wash ddH_2O being removed reaches a pH = 7.0 (usually two to three washes).

9. After the water has been completely removed, transfer ether phase into a round-bottomed glass flask. Add a few drops of ethanol.

10. Evaporate under vacuum using a rotary evaporator in a warm water bath (37°C) until sample is completely dry. Add 1 ml MeOH. Vortex thoroughly (~30 s) while carefully rotating the round-bottomed glass flask.

11. The sample is now ready for analysis. Store at $-20°C$ in a light-impermeable glass test tube.

2.   Solid-phase extraction

1. Activate a C18 cartridge (Sep-Pak) by pushing 20 ml MeOH followed by 20 ml ddH_2O (dropwise) through the packing with a glass syringe.

2. After sample has been collected in the round-bottomed flask, rotoevaporate to dryness.

3. Release vacuum and immediately add 200 μl MeOH. Vortex well. Add 5 ml ddH_2O and vortex. Remove sample to the glass syringe.

4. Repeat step 3.

5. Acidify sample in the glass syringe with a few drops of HCl (0.1 *N*) to a pH ~3.5. Load sample onto activated column.

6. Slowly push 10 ml ddH$_2$O through the column to neutralize pH (~2 drop/s). Check the pH of the eluate using pH paper.

7. Push 10 ml hexane through the column. Collect this fraction with a clean borosilicate glass test tube. Phospholipids elute here.

8. Push 8 ml methyl formate through the column. Collect this fraction. Mono-HETEs, most leukotrienes, prostaglandins, and lipoxins elute here.

9. Push 8 ml MeOH through the column. Collect this fraction. Cysteinyl-leukotrienes elute here.

D. Identification of Human Lipoxygenases and Their Products

Redundant pathways are often available for either LT and LX biosynthesis during multicellular host responses, such as during inflammation and reperfusion injury. To elucidate operative transcellular pathways of formation and the impact of paracrine and other exogenous factors, lipoxygenase (LO) expression can be determined by readily available molecular techniques and followed *in vivo* after challenge with an exogenous agent or during cell incubations *in vitro*. This section reviews aspects of gene analysis by reverse transcription (RT) polymerase chain reaction (PCR) and Northern and Western blotting of importance in eicosanoid research.

1. Lipoxygenase expression

a. *RT-PCR*

Several kits are commercially available for isolation of RNA from isolated cells or tissue. The authors favor the use of TRIzol reagent (e.g., GibcoBRL, Gaithersburg, MD) and phenolic extraction to remove genomic DNA for greatest RNA purity. If messenger RNA is preferred to total RNA, kits utilizing oligo DT columns are available for rapid isolation. The RNA can also be "cleaned" (although total yield will be somewhat reduced) of genomic DNA contamination by using a DNA extraction kit (e.g., Message Clean Kit, GenHunter Corp., Nashville, TN). An example of an RT-PCR protocol for 15-LO is detailed here for illustrative purposes (as in Reference 11).

Note: *Reagents now available may permit streamlining of the protocol so that the RT-PCR of each sample may proceed in the same tube without interruption or transfer of materials.*

Reverse Transcription (RT)
1. Combine ~500 ng of total cellular RNA, RT buffer, 1 mM deoxynucleotide triphosphates (dNTPs), 0.2 µg poly(T) antisense–coding oligonucleotide primer, 40 units

RNacin, 60 μg acetylated BSA, and 400 units M-MLV reverse transcriptase GibcoBRL, Gaithersburg, MD).

2. Incubate at 25°C for 10 min, followed by 42°C for 60 min.

Note: Be certain temperatures are exact.

3. Terminate the reaction by heating to 100°C for 10 min to inactivate the enzyme.
4. Store samples at 4°C until use.

Preparation of Oligonucleotide Primers

Primers are selected to minimize the extensive sequence homologies between 5-, 12-, and 15-LO. Commercial synthesis is now available at minimal cost. Suggested primers for the LOs and related enzymes are listed in Table 3.

Polymerase Chain Reaction (PCR)

1. In 50 μl, combine sense and antisense primers (0.3 μg each), dNTPs (500 μM), DMSO (5%), PCR buffer, *Thermus aquaticus* DNA polymerase (2.5 units), and 5 μl of each RT mixture to act as a template.
2. Overlay samples with two drops of mineral oil.

Note: This step may be omitted if using a heated-cover thermal cycler.

3. Amplify for 25 to 40 cycles in a thermal cycler, with each cycle consisting of
   a. Denaturation: 96°C for 1 min,
   b. Primer annealing: 55°C for 2 min,
   c. Extension: 72°C for 2.5 min.
4. After the last cycle, a longer period of extension (e.g., 5 min) is recommended to even the ends of amplified DNA, especially when PCR products >1 kB are anticipated.
5. Cool to 4°C to terminate the reaction.
6. To determine PCR products, electrophorese 25 μl through a 2% (w/v) agarose gel in TBE buffer containing ethidium bromide (0.5 μg/ml).
7. Photograph with ultraviolet (UV) illumination.

Note: Alternatively, fluorescent product can be detected with a UV transilluminator gel documentation system such as the Gel Doc 1000, Bio-Rad Laboratories, Hercules, CA.

Note: The use of restriction endonucleases with unique targets distinguishing between the homologous LO gene products is strongly recommended. For example, the suggested primers for 15-LO (Table 3) result in a 300 bp PCR product. Incubate the amplified DNA (25 μl) with Apa-1 or Pst-1 (42°C, 16 h) to distinguish, by cleavage of the product, between 15-LO and 5-LO, respectively. Electrophoresis of these reactions through a 3% (w/v) agarose gel in TBE buffer containing ethidium bromide (0.5 μg/ml) should reveal cleavage with only Apa-1.

TABLE 3
Oligonucleotide Primers for Mammalian Lipoxygenases, Cyclooxygenases, and Leukotriene Biosynthetic Enzymes

Gene	Forward Primer	Reverse Primer[a]
5-LO	ATC AGG ACG TTC ACG GCC AGG	CCA GGA ACA GCT CGT TTT CCT G
12-LO	TGG ACA CTG AAG GCA GGG GCT	GGC TGG GAG GCT GAA TCT GGA
15-LO	ATG GGT CTC TAC CGC ATC CGC GTG TCC ACT	CAC CCA GCG GTA ACA AGG GAA CCT GAC CTC
COX-1	CTC ATA GGG GAG ACC ATC AAG	CCT TCT CTC CTA CGA GCT CCT G
COX-2	GCT GAC TAT GGC TAC AAA AGC AGC TGG	ATG CTC AGG GAC TTG AGG AGG GTA
LTA$_4$ hydrolase	GAT GAC TGG AAG GAT TTC C	CCA CTT GGA TTG AAT GCA GAG C
LTC$_4$ synthase	TCC ATT CTG AAG CCA AAG GC	GTG ACA GCA GCC AGT AGA GC
FLAP[b]	GGC CAT CGT CAC CCT CAT CAG CG	GCC AGC AAC GGA CAT GAG GAA CAG G

[a] All primers are listed 5′ to 3′. Sequence information is available on GenBank and in References 11, and 13 to 16.

[b] 5-lipoxygenases activating protein (FLAP).

b. Northern Blot

To better determine RNA levels, a Northern blot can be prepared as in Reference 11.

1. Prepare a 1% (final concentration) RNA agarose gel by mixing 1.5 g agarose with 120 ml DEPC-H_2O. Warm in a microwave until gently boiling (40 to 240 s) to solubilize. Adjust volume for gel box size.

2. Cool in a fume hood to ~60 to 65°C and add 15 ml SEED (10×) and 15 ml formaldehyde.

3. Mix quickly and pour into gel box. (*Note: Avoid bubbles.*)

4. Once hardened, submerge gel in SEED (1×).

5. Prepare RNA to load:

    a. Add 5 to 10 μg total RNA in 10 μl DEPC-H_2O to formamide (7 μl), SEED (10×) (2 μl), formaldehyde (1.2 μl), and ethidium bromide (100 μg/ml) (2 μl).

    b. Incubate RNA mix for 10 to 15 min at 65°C (to avoid secondary structures that would impair electrophoretic mobility).

    c. Place on ice. Quick spin in a microcentrifuge to pool mix at the bottom of the tube.

    d. Load onto gel.

6. Run at 80 to 150 V for 1 to 2 h until RNA has migrated through at least two thirds of the gel.

7. Transfer RNA to a nylon filter (e.g., Amersham Life Science, Arlington Heights, IL) per manufacturer's recommendations (e.g., Vacugene Transfer apparatus, Pharmacia Biotech, Piscataway, NJ).

Radioactive Probe

1. Use ~50 ng cDNA template or PCR product (see above) in 30 μl H_2O.

2. Denature at 100°C, 5 min, and then chill on ice.

3. Spin down (top speed, 30 s) denatured DNA and add:

    10 μl buffer (5×),

    2 μl dNTPs 500 μM (final concentration) (dATP, dGTP, and dTTP),

    2 μl BSA (10 ng/ml),

    1 μl Klenow fragment (5 U/μl).

4. Last, add 5 μl α-$^{32}PO_4$-dCTP (50 μCi).

5. Mix well and incubate 2 to 4 h, 25°C (or 1 h, 37°C).

6. Denature again (100°C, 5 min); then place on ice.

Northern Blot Hybridization

1. To a hybridization jar, add in sequence: 0.1% lauryl sulfate (SDS) (final concentration), H_2O (if needed) to adjust dilutions, 50% formamide, and 5× SSC (final concentration; prepare 20× stock).

2. Carefully place Northern blot into hybridization jar.

3. Add radioactive cDNA probe and incubate overnight at 42°C in a rotating hybridization oven (e.g., Thomas Scientific, Swedesboro, NJ).

4. Wash twice in 2× SSC with 0.1% SDS for 15 min at 25°C.

5. Wash once in 0.1× SSC with 0.1% SDS for 15 min at 25°C.

6. Wash once in 0.1× SSC with 0.1% SDS for 5 min at 65°C, for increased hybridization stringency.

Note: *Decrease the concentration of SSC or increase the temperature to* increase *blot stringency (e.g., 0.2×SSC at 68°C). To determine hybridization of ^{32}P-DNA to blot, expose overnight to either phosphoimager (25°C) or X-ray film with an intensifying screen (–80°C). These methods will develop a visual image of bound radioactivity that permits densitometric analysis.*

Stripping the Blot

1. Incubate 1 h, 65°C in 75% formamide with 2× SSC.

2. Shake periodically.

3. Rinse with 0.1× SSC. Place in either plastic wrap or a sealed bag to keep blot moist. Store at room temperature.

c. Western Blot

To determine if observed enzymatic activity or increased RNA levels are associated with an increase in protein levels of the expected enzyme, perform a Western immunoblot.

1. Following cell incubation, resuspend cell pellets in 150 µl of ice-cold lysis buffer.

2. Add 150 µl of 2× sample buffer to each sample. Boil for 7 min.

3. Load samples onto SDS-polyacrylamide gel. Select a percentage of acrylamide between 5 and 20% depending upon the anticipated molecular weight of the protein. For example, 9% acrylamide can be used to visualize proteins from 30 to 150 kDa (see manufacturer's recommendations).

4. Run gel at 15 mA (150 Vmax) until bromphenol blue in samples reaches the bottom of the gel (~4 h for a 1-mm-thick 10 × 10 cm gel).

5. Transfer to PVDF membrane following manufacturer's recommendations. Ensure equal protein loading and transfer efficiency by Ponceau Red staining of the membrane.

6. Soak membrane (30 min, 25°C) in Tris-buffered saline (TBS) + 5% (w/v) dried milk to diminish nonspecific antibody (Ab) binding.

7. Add selected Ab (see section on Reagents). Gently rock for 45 min at 25°C.

8. Wash three times in TBS (10 min each, 25°C).

9. Incubate with horseradish peroxidase–linked species-specific secondary Ab (dilution 1:15,000 or as specified by manufacturer's protocol).

10. Bound Ab is visualized with enhanced chemiluminescence (ECL) reagent (e.g., Amersham Life Science, Arlington Heights, IL) following manufacturer's protocol.

2.   Product analysis

a. *Radiolabeling*

Labeling arachidonate stores of a single cell type (with radioactivity for HPLC or deuterium for MS) facilitates the tracking of transcellular pathways of formation for eicosanoid products generated during cell–cell interactions. LT and LX production may be complex, involving the bidirectional exchange of biosynthetic intermediates, with rates that are difficult to elucidate without labeling endogenous arachidonate stores. A specific method for labeling platelets with [³H]arachidonate is detailed here to exemplify the technique, but can be generalized for use with other cell types.

Preparation of [³H]arachidonate (C20:4) sodium salt

1.   Obtain [³H]-C20:4 from commercial vendors as a free acid (specific activity 60 to 100 Ci/mmol). Bring ~50 μCi to dryness under a stream of nitrogen in a borosilicate glass tube. Wash down sides of the tube with 1 volume (i.e., the same amount as the starting volume of [³H]-C20:4) of hexane. Once again, bring to dryness under a stream of N_2 while ensuring that the [³H]-C20:4 remains in the bottom of the tube.

2.   Add 50 μl of 0.01 M Na_2CO_3 and 150 μl deionized water (after bubbling each with a stream of N_2 to remove any dissolved O_2). Agitate intermittently for 15 min at 25°C.

3.   Add 3 ml stock buffer and count 0.1% of the total by scintillation spectrometry to determine the total activity of [³H]-C20:4 Na salt.

Incorporating [³H]-C20:4 into Platelet Phospholipids

1.   Add the 3 ml of [³H]-C20:4 Na salt to a suspension of human platelets (7 to 9 × 10⁹ at a dilution of 0.2×10^9/ml) (see Section II.A.2 for platelet isolation).

2.   Incubate 45 min at 37°C with gentle shaking.

3.   Chill on ice for 10 min; then centrifuge at 1450 *g* for 15 min at 4°C.

4.   Reserve supernatant (S_{1450}) and wash the platelets with HEPES-Tyrode buffer.

5.   Repeat centrifugation and pool the supernatant with the S_{1450}. Resuspend platelets in reaction buffer and store at 4°C.

6.   Determine percentage [³H]-C20:4 incorporation by scintillation spectrometry of an aliquot from the pooled supernatants (unincorporated label) and platelet suspension (incorporated label). Percent incorporation should be 50 to 80%. This value varies depending on donor variables and the state of cellular activation.

b. *Physical Methods*

RP-HPLC separates compounds on the basis of size, charge, and hydrophobicity that are present in an extract of materials from cellular incubations. High pressure is used to lessen the time for sample analysis. Eluting materials can be detected in the picogram range by coupling the separation achieved by HPLC to a diode array detector (DAD), electrochemical detector, refractive index measurement, or mass spectrometer. Spectrophotometric analysis is most commonly employed because most eicosanoids give distinctive UV chromophores. For example, LTs in methanol have a conjugated triene with λmax ~ 270 nm for LTB$_4$ and DiHETEs and λmax ~ 280 nm for cysteinyl LTs, while both LX and epimer-LX have a conjugated tetraene with λmax ~ 300 nm (Figure 4). Although PGB$_2$ carries a UV chromophore with λmax ~ 270 nm, most prostanoids absorb UV at λmax < 210 nm

Note: *PGB$_2$ is not a product of PLT-PMN coincubation and therefore is ideally suited for use as an internal standard in these coincubations.*

Because numerous interfering substances absorb UV at <210 nm, the detection of most prostanoids solely on the basis of UV criteria is limited. To avoid this problem, couple RP-HPLC (to separate PGs with (1) immunoassay for specific PG measurement, or with (2) electrochemical detection or (3) electrospray MS to profile PG reaction products and thereby achieve sensitive PG identification and quantitation.

The length of time after injection for a compound to reach the detector is defined as its retention time. The retention time combined with other physical properties of the compound can be used to identify individual eicosanoids when synthetic standards are available. Retention time and spectrophotometric analysis via DADs permit online identification and quantitation of LT and LX. When available, coelution with authentic standards by matching studies (i.e., by injecting 50% sample and 50% standard to determine if the retention times and UV spectra of the sample match the standard) can confirm the identification of unknown compound(s) present in a biological matrix.

Figure 4 is an example of an RP-HPLC chromatogram of eicosanoid products from stimulated PMN (Figure 4A) and PLT (Figure 4B). RP-HPLC was performed using a Hewlett Packard 1100 Series Diode Array Detector equipped with a binary pump and eluted on a LUNA C18-2 microbore column (150 × 1 mm, 5 μm) (Phenomenex, Torrance, CA) using a mobile phase composed of methanol/water/acetate (58/42/0.01, v/v/v) as phase 1 (0 to 25 min) and a linear gradient with methanol/acetate (99.99/0.01, v/v) as phase 2 (25 to 37 min) at a flow rate of 0.12 ml/min. Collected UV data were recalled at 270 nm (Figure 4A) to detect conjugated trienes (e.g., leukotrienes) and 301 nm (Figure 4B) to detect conjugated tetraenes (e.g., lipoxins).

Also of great utility in the sensitive detection of LT and LX biosynthesis is MS which consists of three basic components: an ion source, a mass analyzer, and a detector. Several different techniques for generating ions from compounds in solid, liquid, or gaseous phases are available with current mass spectrometers; however, consideration of this area is beyond the scope of this chapter. Mass analyzers utilize electromagnetic fields to separate molecular ions on the basis of their mass to charge ratio (*m/z*) which enables detection of unique molecular fragments. Separation of materials entering the MS is required when analyzing individual compounds from a biological matrix and can be carried out by gas chromatography (GC) using an inert carrier gas (e.g., helium) or by liquid chromatography (LC) via capillary flow, HPLC, or direct injection. MS has the advantage of detection in the same low picogram range as UV yet does not require the presence of a chromophore. In addition, structural elucidation of unknown compounds is possible by MS via analysis of its molecular ion, fragmentation pattern (i.e., presence of prominent ions), and changes in the sizes of these fragments after derivatization (e.g., by silylation,

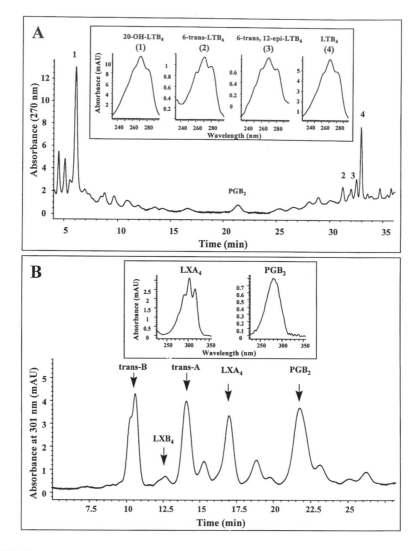

FIGURE 4

Transcellular biosynthesis of lipoxins during neutrophil and platelet interactions. (A) Representative RP-HPLC chromatogram of endogenous products obtained from human neutrophils exposed to lipopolysaccharide (LPS) (1 µg/ml) and stimulated with platelet-activating factor (PAF) (300 nM) and fMLP (100 nM). Collected UV data was recalled at 270 nm. UV chromophores of endogenous 20-OH-LTB$_4$, LTB$_4$, and the nonenzymatic hydrolysis products of the epoxide LTA$_4$, denoted compound I (6-*trans*-LTB$_4$) and compound II (6-*trans*-12-epi-LTB$_4$) are shown in the insets. (B) Representative RP-HPLC chromatogram of products obtained from human platelets incubated (20 min, 37°C) with LTA$_4$ (20 µM) and the calcium ionophore, A23187 (5 µM). *trans*-B denotes the retention time for the 8-*trans*-LXB$_4$ and 14-*S*-8-*trans*-LXB$_4$ isomers, and *trans*-A indicates the retention time for the 11-*trans*-LXA$_4$ and 6-*S*-11-*trans*-LXA$_4$ isomers. Collected UV data were recalled at 301 nm. The specific UV chromophore of LTA$_4$-derived LXA$_4$ and the internal standard, PGB$_2$, are shown in the insets. Vertical arrows indicate the retention time of authentic standards.

ozonolysis, or catalytic hydrogenation). Drawbacks of GC/MS include a requirement for compounds to be volatile to be suitable for GC and significant losses of material during derivitization reactions.

Newer methods utilize electrospray ionization MS coupled to HPLC to enable analysis of individual compounds directly after elution from the HPLC column. LC/MS/MS is the most powerful technique currently available for eicosanoid analysis with detection limits in the low picogram range.

Figure 5 is an example of LC/MS/MS analysis of lipoxin products obtained from human PLT incubated with LTA_4, a pivotal biosynthetic intermediate for either cysteinyl LT or LX (see Figures 1 and 2). Eicosanoids were extracted from the incubation mixture by solid-phase extraction, and materials in the methyl formate elution were injected into a SpectraSYSTEM HPLC (Thermo Separation Products, San Jose, CA) coupled via an electrospray ionization source to an LCQ quadrupole ion trap MS (Finnigan MAT, San Jose, CA). A LUNA C18-2 (150×2 mm, 5 µm) column was eluted isocratically with methanol/water/acetic acid (58:42:0.009, v/v/v) at 0.2 ml/min into the electrospray probe. The LCQ spray voltage was set to 6 kV and the heated capillary to -4 V and 250°C. Full-scan MS were acquired by scanning between m/z 340 and 360 in the negative ion mode, followed by the acquisition of product ion mass spectra (MS/MS) for m/z 351.5 ([M-H]$^-$ of lipoxin A_4 and lipoxin B_4). The trace in Figure 5A is the UV chromatogram monitored at 300 nm and shows many peaks. In sharp contrast, the greatly simplified traces in Figure 5B and C are selected ion monitoring (SIM) chromatograms of MS/MS daughter ions unique to LXA_4 and LXB_4, respectively (m/z 235 for LXA_4 and m/z 221 for LXB_4). Inset are the MS/MS spectra for LXA_4 and LXB_4, further confirming the identity of the products.

c. Immunoassay

Enzyme-based and radioactive immunoassays are available for a wide variety of eicosanoids, including specific LT and LX. This method also detects LT and LX in the picogram range and is an even more convenient and translatable method for analyzing large numbers of samples. Several caveats should be noted: (1) despite using an Ab with high affinity for the compound of interest, nonspecific Ab crossreactivity remains a frequent problem; (2) the presence of interfering substances that inhibit antigen–Ab interaction are common in biological matrices; (3) analysis is limited to a specific compound; and (4) consistent and accurate pipetting is essential to acquiring reproducible results. These common problems plus the lack of structural information necessitate initial validation of the selected immunoassay with physical chemical techniques, running appropriate controls and performing at least two to three determinations for each sample. Here an LXA_4 enzyme-linked immunosorbent assay (ELISA) is described to exemplify a protocol that is, in general, similar to those used for other eicosanoid ELISAs of interest (e.g., TxB_2, LTB_4, and LTC_4).

LXA_4 ELISA

LXA_4 antisera was prepared by conjugation to keyhole limpet hemocyanin through the succinimide ester of LXA_4. Rabbits were immunized initially with

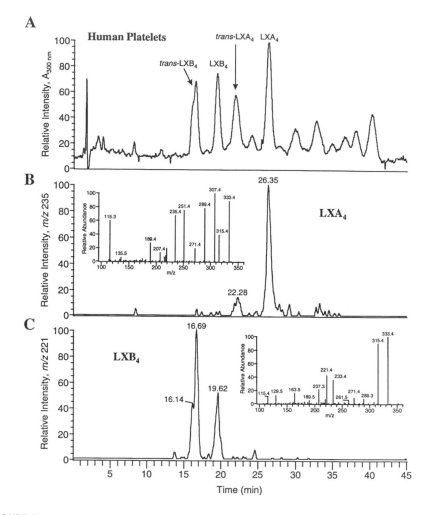

FIGURE 5

LC/MS/MS profile of products obtained from human platelets incubated with LTA$_4$. (A) UV chromatogram of eicosanoids monitored at 300 nm. *trans*-LXB$_4$ denotes retention times of 8-*trans*-LXB$_4$ and 14(*S*)-8-*trans*-LXB$_4$, and *trans*-LXA$_4$ denotes retention times of 11-*trans*-LXA$_4$ and 6(*S*)-11-*trans*-LXA$_4$. (B) Selected ion monitoring (SIM) chromatogram of unique LXA$_4$ *m/z* 235.5 daughter ion derived from MS/MS of *m/z* 351.5 parent ion ([M-H]$^-$ of lipoxins A$_4$ and B$_4$). Inset, MS/MS spectrum shows prominent diagnostic LXA$_4$ daughter ions at *m/z* 333 [351 - H$_2$O], 315 [351 - 2H$_2$O], 307 [351 - CO$_2$], 289 [351 - H$_2$O, - CO$_2$], 271 [351 - 2H$_2$O, - CO$_2$], 251 [351 - CHO(CH$_2$)$_4$CH$_3$], 235 [351 - CHO(CH$_2$)$_3$COOH], 207 [351 - CO$_2$, - CHO(CH$_2$)$_4$CH$_3$], 189 [351 - H$_2$O, - CO$_2$, - CHO(CH$_2$)$_4$CH$_3$], 135 [351 - CHO(CH$_2$)$_3$COOH, - CHO(CH$_2$)$_4$CH$_3$], and 115 [CHO(CH$_2$)$_3$COO$^-$]. (C) SIM chromatogram of unique LXB$_4$ *m/z* 221.5 daughter ion derived from MS/MS of *m/z* 351.5 parent ion. Inset, MS/MS spectrum shows prominent diagnostic LXB$_4$ daughter ions at *m/z* 333 [351 - H$_2$O], 315 [351 - 2H$_2$O], 289 [351 - H$_2$O, - CO$_2$], 271 [351 - 2H$_2$O, - CO$_2$], 233 [351 - H$_2$O, - CHO(CH$_2$)$_4$CH$_3$], 221 [351 - CHOCHOH(CH$_2$)$_4$CH$_3$], 207 [351 - CO$_2$, - CHO(CH$_2$)$_4$CH$_3$], 189 [351 - H$_2$O, - CO$_2$, - CHO(CH$_2$)$_4$CH$_3$], 163 [351 - CO$_2$, - CH$_2$COHCHOH(CH$_2$)$_4$CH$_3$], 129 [CH$_3$CO(CH$_2$)$_3$COO$^-$], and 115 [CHO(CH$_2$)$_3$COO$^-$].

100 μg of the conjugate per rabbit. Monthly booster injections with 50 μg of the initial conjugate were performed and rabbits were bled through ear veins. Antisera was collected from the blood by centrifugation (900 g), stored at −20°C and diluted 1:200 in ELISA buffer (optimal dilutions for the assay conditions).

Note: *Horseradish peroxidase–labeled LXA_4 was also prepared through its succinimide ester.[12]*

1. Precoat a 96-well microtiter plate with 1 μg affinity-purified goat antirabbit IgG per well (or purchase precoated plate).
2. Prepare serial dilutions of LXA_4 (20 to 2000 pg/ml ELISA buffer).
3. In sequence, add 50 μl LXA_4 antiserum (diluted 1:200 in ELISA buffer), then 50 μl LXA_4 standard or sample in duplicate, and finally 50 μl horseradish peroxidase–labeled LXA_4.

Note: *Use a repeating pipettor whenever possible to minimize systematic pipetting error.*

4. Gently shake plates for 1 h at 25°C.
5. Wash three times with 200 μl ELISA wash buffer per well. (Use multichannel repeating pipettor.)
6. Tap plate on absorbent diaper until no visible liquid remains in the wells.
7. Add 150 μl K-Blue® substrate and incubate up to 15 min at 25°C.
8. Stop reaction with 100 μl of 1 N H_2SO_4.
9. Read absorbance at 450 nm. Quantitate amount of LXA_4 in samples by comparison with the LXA_4 serial dilutions.

Note: *A standard curve must be included in each 96-well microtiter plate for reference.*

Figure 6 demonstrates the utility of ELISAs when measuring LT and LX in biological samples. Levels of immunoreactive LTC_4 and LXA_4 were determined in nasal lavage fluid from aspirin-sensitive patients with asthma both before and after a placebo or threshold dose of oral aspirin that elicited an *in vivo* inflammatory response in patient airways (as in Reference 12).

Reagents Needed

A. Buffers and solutions

Acid Citrate Dextrose (ACD)

Citric acid (Sigma C-0759), 38 mM
Trisodium citrate (Sigma C-7254), 75 mM
Dextrose (Sigma G-8870), 136 mM

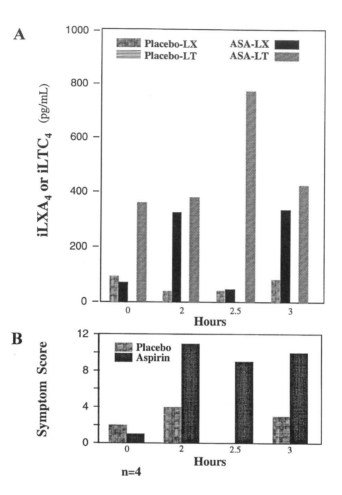

FIGURE 6

Eicosanoids present in nasal lavage fluid from patients with aspirin-sensitive asthma. In a randomized, placebo-controlled double-blind investigation, volunteer subjects with aspirin-sensitive asthma were lavaged with 5 ml of 0.9% saline 2 to 3 h after ingestion of either a placebo or previously determined threshold dose of aspirin (sufficient to generate ocular, nasal, and/or respiratory symptoms). (A) Levels of immunoreactive LTC_4 or LXA_4 were determined (mean, n = 4, d = 2) and correlated with patient symptom score (B).

HEPES-Tyrode

Sodium chloride (Sigma S-7653), 128 mM
Sodium bicarbonate (Sigma S-6297), 8.9 mM
Potassium chloride (Sigma P-9333), 2.7 mM
Potassium phosphate, monobasic (Sigma P-5379), 0.8 mM
Dextrose, 5.5 mM
HEPES (Sigma H-7523), 9.7 mM
*Titrate pH to 7.4, then add:
Magnesium chloride (Sigma M-2670), 1.0 mM

Dulbecco's Phosphate Buffered Saline (PBS) (BIOWHITAKER No. 17-512F)

Potassium chloride, 2.7 mM
Potassium phosphate, monobasic, 1.5 mM
Sodium phosphate, dibasic (Sigma S-7907), 8.1 mM
Sodium phosphate, heptahydrate (Sigma S-9390), 8.1 mM
Sodium chloride, 137 mM
*Adjust pH to 7.4

Hanks' Balanced Salt Solution (HBSS) (10×) (Gibco No. 14180-012)

Potassium chloride, 53.7 mM
Potassium phosphate, monobasic, 4.4 mM
Sodium chloride, 1.37 M
Sodium phosphate, heptahydrate, 3.4 mM
Dextrose, 55.5 mM

DEPC-H$_2$O

Diethyl pyrocarbonate (DEPC) (Sigma D-5758), 1 ml/l dH$_2$O
*Shake, let stand overnight. Autoclave for 30 min.

TBE

Trizma base (Sigma T-6791), 100 mM
Boric acid (Sigma B-7660), 90 mM
EDTA (Sigma E-9884), 1 mM
*Adjust pH to 8.4. Autoclave for 30 min.

SEED (10×)

Sodium acetate (Sigma S-7670), 80 mM
3-[N-Morpholino]propanesulfonic acid (MOPS)
(Sigma M-5162), 200 mM
EDTA, 10 mM
*Prepare working solutions of sodium acetate (3 M) and EDTA
(0.5 M, pH 8.0) and dilute to final volume with dH$_2$O. Adjust pH
to 7.0 with a few drops of 10 N NaOH. Filter with 0.2 μM membrane
and store in a dark bottle. Do *not* autoclave.

SSC (20×)

Sodium chloride, 3.0 M
Sodium citrate, 300 mM
*Adjust pH to 7.0 with a few drops of 10 N NaOH. Autoclave for 30
min.

Reverse Transcription (RT) Buffer

Purchase with reverse transcriptase

Polymerase Chain Reaction (PCR) Buffer

Purchase with DNA polymerase

Tris-Buffered Saline (TBS)

Trizma HCl (Sigma T-6666), 25 mM
Sodium chloride, 200 mM
Tween 20 (Sigma P-7949), 0.15%
*Adjust pH to 7.6 with a few drops of 10 N NaOH

Cell Lysis Buffer for Western Blotting

HEPES, 10 mM
Calcium chloride (Sigma C-5080), 1.6 mM
Triton X-100 (Sigma T-9284), 0.5%
Phenylmethylsulfonyl fluoride (Sigma P-7626), 10 µg/ml
Leupeptin (Sigma L-2023), 10 µg/ml
Aprotinin (Sigma A-6279), 10 µg/ml
*Add reagents to HBSS and adjust pH to 7.4

Sample Buffer for Western Blotting (2×)

Trizma HCl, 125 mM
Lauryl sulfate (SDS) (Sigma L-4509), 8%
β-Mercaptoethanol (Sigma M-6250), 10%
Glycerol (Sigma G-6279), 17%
Leupeptin, 20 µg/ml
Aprotinin, 20 µg/ml
Bromphenol blue (Sigma B-0126), 0.05%

ELISA Buffer

Sodium chloride, 0.9%
Potassium phosphate, monobasic, 100 mM
Potassium phosphate, dibasic (Sigma P-5504), 100 mM
Bovine serum albumin (Sigma A-2153), 0.1%
*Adjust pH to 7.4

ELISA Wash Buffer

Potassium phosphate, monobasic, 10 mM
Potassium phosphate, dibasic, 10 mM
Tween 20, 0.05%
*Adjust pH to 7.4

K-Blue Substrate

Purchase with ELISA kit or from Neogen No. 300175

B. Compounds, agonists, and cytokines

For cell isolation, dextran and LSM are purchased from Sigma Chemical Co. (St. Louis, MO) and Organon Teknika Corp. (Durham, NC), respectively.

A variety of sources are available for most of the major eicosanoids, including LT, LX, and some of their biosynthetic intermediates. Consider Biomol (Plymouth meeting, PA), Calbiochem (San Diego, CA), Cayman Chemical (Ann Arbor, MI), Oxford Biomedical Research (Oxford, MI) and, for radioactive eicosanoids, American Radiolabeled Chemicals, Inc. (St. Louis, MO), Amersham Life Sciences (Arlington Heights, IL), and New England Nuclear (Boston, MA).

Note: *It is essential to verify the concentration and integrity of all commercial eicosanoid stocks by at least UV absorbance as a routine laboratory procedure prior to use and, if possible, determine the HPLC profile for the compound to assess its purity and the presence of isomerization products.*

Cell agonists and cytokines are also available from a large number of vendors, including Boehringer Mannheim (Indianapolis, IN), Endogen (Woburn, MA), R&D Systems (Minneapolis, MN), and Sigma Chemical Co. (St. Louis, MO). Follow the suppliers' recommendations carefully for the optimal vehicle for agonist or cytokine. Most are soluble in ethanol, which can be safely used for cell incubations at concentrations <0.4% if proper vehicle controls are run in parallel. Some agonists are more soluble in DMSO (e.g., A23187) or water (e.g., thrombin). In general, prepare these agents 500 to 1000-fold more concentrated than the desired final concentration so that they may be added in a small amount of vehicle (e.g., 1 to 5 μl). Further dilute each agonist in buffer (e.g., 25 to 50 μl) just prior to addition to cells so that the local concentration of delivered vehicle is not inordinately high.

C. Inhibitors

Inhibitors of LT and LX biosynthetic enzymes (Table 2) are available from suppliers (see above). Most are stable at −20°C in an organic solvent. Take special precaution with aspirin, which must be prepared just prior to its addition.

D. Molecular biology reagents

Reagents for RNA isolation, RT-PCR (e.g., dNTPs and DNA polymerases), Northern and Western blotting (e.g., Ponceau Red stain) are widely available. Of special interest is the recent availability of specific antibodies directed against COX-1 (Oxford Biomedical Research), COX-2 (Biomol, Calbiochem), 5-LO (Cayman Chemical), 12-LO (Oxford Biomedical Research), and the LTB_4 receptor (polyclonal, Cayman Chemical).

III. Discussion

Detection and determination of LT and LX biosynthesis during multicellular responses can be complex and require a highly organized experimental approach.

For the investigator new to this area, methods described here have outlined techniques for the isolation of cells from whole blood, establishment of relevant incubation conditions, and determination of operative transcellular pathways for LT and LX during cell–cell interactions.

Platelet–neutrophil interactions occur *in vivo* at sites of vascular inflammation, are well characterized, and were used here to exemplify practical aspects of studying LT and LX transcellular biosynthesis. Cell–cell contact facilitates the exchange of eicosanoid intermediates, yet is not required for transcellular eicosanoid biosynthesis. The concepts presented here can be applied much more broadly to many other scenarios where circulating cells interact with tissue-resident cells during multicellular host responses.

The study of transcellular biosynthesis of LT and LX is in its infancy. The authors trust that the methodologies presented here will assist others in their exploration of this emerging and exciting area of investigation.

Acknowledgments

Work described from the Serhan laboratory was supported in part by the NIH (DK #50305 and GM #38765) to CNS. BDL is a receipient of a mentored Clinical Scientist Development Award from the National Institutes of Health (NHLBI-K08-HL03788). KG is a recipient of a postdoctoral fellowship from the National Arthritis Foundation. The authors thank Marc Pouliot and Nan Chiang for their helpful comments. To our colleagues, the references provided are not exhaustive, but rather meant to exemplify technical points in the investigation of eicosanoid transcellular biosynthesis.

References

1. **Serhan, C. N., Haeggstrom, J. Z., and Leslie, C. C.,** Lipid mediator networks in cell signaling: update and impact of cytokines, *FASEB J.*, 10, 1147–1158, 1996.
2. **Samuelsson, B., Dahlen, S. E., Lindgren, J. A., Rouzer, C. A., and Serhan, C. N.,** Leukotrienes and lipoxins: structures, biosynthesis, and biological effects, *Science*, 237, 1171–1176, 1987.
3. **Serhan, C. N.,** Lipoxins and novel aspirin-triggered 15-epi-lipoxins: a jungle of cell–cell interactions or a therapeutic opportunity? *Prostaglandins*, 53, 107–137, 1997.
4. **Fiore, S. and Serhan, C. N.,** Formation of lipoxins and leukotrienes during receptor-mediated interactions of human platelets and recombinant human granulocyte/macrophage colony-stimulating factor-primed neutrophils, *J. Exp. Med.*, 172, 1451–1457, 1990.
5. **Funk, C.,** The molecular biology of mammalian lipoxygenases and the quest for eicosanoid functions using lipoxygenase-deficient mice, *Biochim. Biophys. Acta*, 1304, 65–84, 1996.
6. **Bigby, T. D., Hodulik, C. R., Arden, K. C., and Fu, L.,** Molecular cloning of the human LTC_4 synthase gene and assignment to chromosome 5q35, *Mol. Med.*, 2, 637–646, 1996.

7. **Yokomizo, T., Izumi, T., Chang, K., Takuwa, Y., and Shimizu, T.,** A G-protein-coupled receptor for leukotriene B_4 that mediates chemotaxis, *Nature*, 387, 620–624, 1997.

8. **Lewis, R. A., Austen, K. F., and Soberman, R. J.,** Leukotrienes and other products of the 5-lipoxygenase pathway. Biochemistry and relation to pathobiology in human diseases, *N. Engl. J. Med.*, 323, 645–655, 1990.

9. **Irvin, C. G., Tu, Y. P., Sheller, J. R., and Funk, C. D.,** 5-Lipoxygenase products are necessary for ovalbumin-induced airway responsiveness in mice, *Am. J. Physiol.*, 272, L1053–L1058, 1997.

10. **Shen, J., Herderick, E., Cornhill, J. F., Zsigmond, E., Kim, H.-S., Kuhn, H., Guervara, N. V., and Chan, L.,** Macrophage-mediated 15-lipoxygenase expression protects against atherosclerosis development, *J. Clin. Invest.*, 98, 2201–2208, 1996.

11. **Levy, B. D., Romano, M., Chapman, H. A., Reilly, J. J., Drazen, J. M., and Serhan, C. N.,** Human alveolar macrophages have 15-lipoxygenase and generate 15(S)-hydroxy-5,8,11-*cis*-13-*trans*-eicosatetraenoic acid and lipoxins, *J. Clin. Invest.*, 92, 1572–1579, 1993.

12. **Levy, B. D., Bertram, S., Tai, H. H., Israel, E., Fischer, A., Drazen, J. M., and Serhan, C. N.,** Agonist-induced lipoxin A_4 generation: detection by a novel lipoxin A_4-ELISA, *Lipids*, 28, 1047–1053, 1993.

13. **Funk, C. D. and FitzGerald, G. A.,** Eicosanoid forming enzyme mRNA in human tissues, *J. Biol. Chem.*, 266, 12508–12513, 1991.

14. **Hla, T. and Neilson, K.,** Human cyclooxygenase-2 cDNA, *Proc. Natl. Acad. Sci. U.S.A.*, 89, 7384–7388, 1992.

15. **Sanak, M., Simon, H.-U., and Szczeklik, A.,** Leukotriene C_4 synthase promoter polymorphism and risk of aspirin-induced asthma, *Lancet*, 350, 1599–1600, 1997.

16. **Jakobsson, P.-J., Shaskin, P., Larsson, P., Feltenmark, S., Odlander, B., Aguilar-Santelises, M., Jondal, M., Biberfeld, P., and Claesson, H.-E.,** Studies on the regulation and localization of 5-lipoxygenase in human B-lymphocytes, *Eur. J. Biochem.*, 232, 37–46, 1995.

Chapter

Purification and High-Resolution Analysis of Anandamide and Other Fatty Acylethanolamides

Andrea Giuffrida and Daniele Piomelli

Contents

I.   Introduction

Anandamide (arachidonylethanolamide) was the first cannabimimetic substance identified in mammalian brain.[1] Like Δ^9-tetrahydrocannabinol, the active principle of *Cannabis*, anandamide binds with high affinity to cannabinoid CB1-type receptors and with somewhat lower affinity to CB2-type receptors. Moreover, it mimics many psychotropic effects typical of cannabimimetic drugs (for review, see Reference 2). Anandamide is released from brain neurons in an activity-dependent manner[3] through a mechanism that involves phosphodiesterase-mediated cleavage of the membrane phospholipid precursor, *N*-arachidonyl phosphatidylethanolamine (PE)[4,5] (Figure 1). Carrier-mediated uptake followed by enzymatic hydrolysis is thought to terminate its biological effects.[6–10]

Additional acylethanolamides (or *N*-acylethanolamines, AEs) with distinctive cannabimimetic and noncannabimimetic activities have been identified,[11] suggesting that anandamide belongs to a broader family of signaling lipids acting not only in brain, but also in peripheral tissues (Figure 2). The possible physiological roles of these compounds are still largely unexplored, although several lines of evidence suggest that they may serve regulatory functions including control of vascular tone, intestinal motility, and immune responses.[12–14]

This chapter describes a series of methodological approaches for the extraction, purification, and analysis of anandamide and related AEs. A flow chart for purification and analysis of AEs is shown in Figure 3. Some general concepts on the theory and practice of the individual techniques are provided in each section, followed by detailed protocols of the procedures as well as discussions of the problems expected to arise at each analytical step. The fractionation and analysis of the AE precursors *N*-acyl PE are considered in a separate section. Sources of chemicals and recipes for the buffers used are given at the end of the chapter.

II.   Protocol

A.   Extraction Techniques

1.   Basic methodological philosophy

Extraction into an organic solvent is generally used as a first step in the analysis of AEs, and has several advantages: it eliminates contaminants, allowing for a subsequent

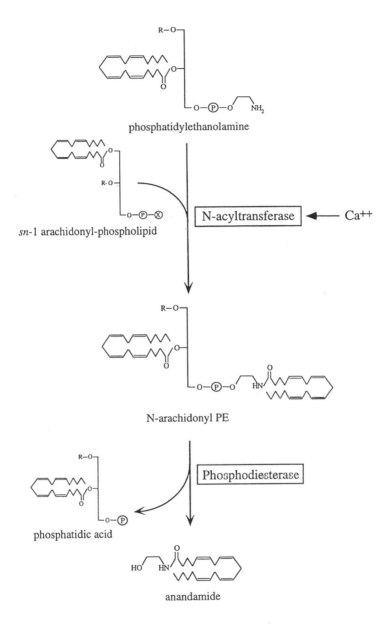

FIGURE 1
Hypothetical model of anandamide formation from *N*-arachidonyl PE in rat brain tissue. The model also shows the enzymatic step involved in the biosynthesis of *N*-arachidonyl PE from phosphatidyle-thanolamine. A rise in the intracellular Ca^{2+} concentration produced by neuronal activity may stimulate *N*-acyltransferase activity, which catalyzes the transfer of arachidonate from *sn*-1 arachidonyl phospholipids to phosphatidylethanolamine.

ACYLETHANOLAMIDES	RECEPTORS
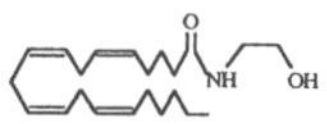 arachidonylethanolamide (20:4 $\Delta^{5,8,11,14}$)	CB1,[1] CB2[2]
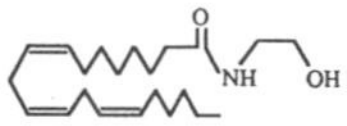 homo-γ-linolenylethanolamide (20:3 $\Delta^{8,11,14}$)	CB1[3,4]
 docosatetraenoylethanolamide (22:4 $\Delta^{7,10,13,16}$)	CB1[3]
 stearylethanolamide (18:0)	?
 oleylethanolamide (18:1 Δ^9)	?
 linoleylethanolamide (18:2 $\Delta^{9,12}$)	?
 palmitylethanolamide (16:0)	CB2-like[2]
 palmitoleylethanolamide (16:1 Δ^9)	?

1. Devane, W. A. et al., Science, 258, 1946–1949, 1992.   3. Hanuš, L. et al., J. Med. Chem., 36, 3032–3034, 1993.
2. Facci, L. et al., Proc. Natl. Acad. Sci. U.S.A., 92, 3376–3380, 1995.   4. Felder, C. C. et al., Proc. Natl. Acad. Sci. U.S.A., 90, 7656–7660, 1993.

FIGURE 2
Chemical structures of the most common acylethanolamides found in mammalian tissues and their selectivity for cannabinoid CB1 and CB2 receptor subtypes.

satisfactory purification; it concentrates material and improves the sensitivity of the analysis. Disadvantages mainly consist in loss of sample, a point that must be carefully considered as AEs are generally present in biological samples at very low concentrations. Also, it is advisable to extract AEs from the tissue as soon as possible after harvesting it, to minimize changes in lipid composition that may occur post-mortem.[4,15,16] If gas chromatography/mass spectrometry (GC/MS) analysis is pursued, all the extraction procedures should be performed in glass tubes to avoid sample contamination by plasticizers. Sylanization of the glassware is not indispensable.

The choice of the particular extraction method is mainly dictated by whether the source of AEs is a tissue or a biological fluid. In either case, a mixture of

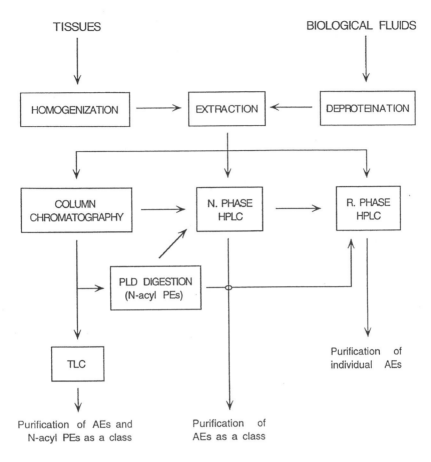

FIGURE 3
Suggested purification strategies for AEs.

methanol/chloroform generally gives satisfactory AE extraction. However, biological fluids containing large amounts of protein (e.g., blood plasma) may produce emulsions and protein precipitation at the methanol–chloroform interface, making extraction difficult and generating erratic yields. The extraction procedure described below should help overcome these problems.

2.   Extraction of acylethanolamides from tissue samples

Procedure

1.   Homogenize fresh tissue at 0 to 4°C in 50 mM Tris buffer, pH 8, containing 0.32 M sucrose, or in a mixture of methanol/Tris buffer (1:1, v/v) to a final 20-fold dilution. Comparable results have been obtained with either method.

2.   Extract AEs in a screw-capped glass vial by vortexing the homogenate vigorously with a mixture of chloroform/methanol/buffer (2:1:1).

3. Recover the chloroform phase (lower layer), evaporate to dryness under a stream of
    N_2 with mild heating, and reconstitute in an appropriate solvent for further processing
    (see below).

Note: *The separation of the aqueous and the organic layer may be improved by
           a brief centrifugation of the sample (800 g for 10 min), or by adding a
           small volume of a saturated NaCl solution. Particulate material in the
           chloroform phase may be eliminated by filtering through 0.45 µm low-
           binding Durapore membranes (Millipore, Bedford, MA).*

3.   Extraction of acylethanolamides from biological fluids

Before lipid extraction, a protein precipitation step is recommended for biological
fluids containing large amounts of protein. The following procedure, which was set
up for rat blood plasma, may be adapted to other biological fluids.[17]

Procedure

1. Collect blood samples (5 ml) using a syringe filled with 2 ml Krebs-Tris buffer
    containing EDTA (4.5 mM).
2. Centrifuge samples are in Accuspin™ tubes (Sigma, St Louis, MO) for 10 min at 18°C
    ($800 \times g$) and carefully recover the plasma layers and dilute them with cold acetone
    (–20°C, 2 volumes) to precipitate plasma protein. Remove the protein by centrifugation
    at $1000 \times g$ for 10 min.
3. Evaporate residual acetone in the supernatant under N_2 to obtain a volume close to the
    initial one. The supernatant is then subjected to lipid extraction using a mixture of
    methanol/chloroform, as described above.

4.   Storage of lipid extracts

Because of the sensitivity of AEs to oxidation, adsorption, and hydrolysis, extracts
should never be stored for extended periods of time. In the authors' experience, best
results are obtained by analyzing freshly prepared material. For a relatively short storage
(1 to 2 weeks), dissolving the extracts in high-grade pure methanol or chloroform and
storing them at –80°C in screw-capped glass vials flushed with N_2 is suggested.

B.   Fractionation and Purification

Column chromatography and thin-layer chromatography (TLC) are among the most
common procedures used to fractionate and purify lipid mixtures. Silica gel G chromatography on small "homemade" columns offers a rapid and inexpensive initial
fractionation of AEs from tissue extracts, allowing for the removal of many potential
contaminants. This step can be omitted and replaced with filtration through 0.45-µm
low-binding Durapore filters (Millipore, Bedford, MA) during the fractionation of
AEs derived from biological fluids.

1.   Column chromatography

Procedure

1.   Prepare a slurry with 1 volume of silica (230 to 400 mesh, 40 to 63 μm, Sigma) and 1 volume of chloroform in a beaker. Gently stir the slurry to homogeneity.

2.   Immediately pour 1 ml of the slurry into a Pasteur pipette plugged with glass wool, and wash the column bed (final volume = 0.5 ml) with 2 ml of chloroform.

3.   Load the chloroform extract containing AEs down the sides of the column. Allow the sample to enter the column; then wash with chloroform (1 ml) to ensure the transfer of the lipid mixture throughout the gel matrix.

4.   Elute the AEs by gravity with a mixture (2 ml) of chloroform/methanol (9:1). The fractions, which also contain other neutral lipids (mainly cholesterol, mono-, di-, and triacylglycerols), are collected and evaporated to dryness.

2.   HPLC techniques

Fractions containing AEs can be subjected to normal- or reverse-phase high performance liquid chromatography (HPLC) for further fractionation and purification. Although time-consuming, the HPLC step before GC/MS analysis is recommended, as it significantly improves GC resolution and prolongs column life.

a. Procedure: Normal-Phase HPLC

1.   Dissolve fractions containing AEs in chloroform (100 μl) and inject then into an HPLC equipped with a normal-phase Resolve Silica column (3.9 mm × 15 cm, 5 μm, Waters, Milford, MA).

2.   Elute the column with a gradient of isopropyl alcohol (B) in *n*-hexane (A) (100% A initial; 90% A, 10% B for 1 min, 60% A, 40% B for 7 min, 50% A, 50% B for 12 min) at a flow rate of 1.7 ml/min. Under these conditions, all AEs are eluted from the HPLC column between 4.7 and 5.3 min.

3.   Collect AE-containing fractions in glass reaction vials (Supelco, Bellefonte, PA) and evaporate the solvent under N_2.

b. Procedure: Reverse-Phase HPLC

Reverse-phase HPLC on C18 columns can be used to resolve individual AEs, based on the length of their acyl chain and their degree of unsaturation. As a rule, the longer and the more saturated the AE acyl chain, the longer the retention time.

1.   Dissolve AE-containing fractions in 50 to 100 μl of chloroform and inject into a HPLC equipped with a Novapak C18 column (4.6 mm × 15 cm, 4 μm, Waters).

2.   Elute the column with a linear gradient of methanol in water (75 to 100% over 50 min), at a flow rate of 1.0 ml/min. Under these chromatographic conditions, AEs differing in one double bond are separated by at least 2.5 min.

Note: *Reverse-phase HPLC on Novapak columns does not allow the resolution of anandamide from γ-linoleoylethanolamide. To achieve this separation, the authors used a Free Fatty Acid HP column (4.6 mm × 15 cm, 5 mm, Waters) eluted under isocratic conditions with a solvent mixture of 70% acetonitrile/tetrahydrofuran/water, 45:20:35 in water, at a flow rate of 1.5 ml/min.[18]*

3.   Procedure: Thin-layer chromatography

TLC on silica gel plates is an inexpensive and versatile technique to fractionate AEs. Good results on a qualitative basis can be rapidly obtained using a minimum of equipment. TLC provides greater resolution than column chromatography, but it is relatively inefficient if quantitation is needed (yields of fractionated AEs after scraping from the plate are usually low, varying between 50 and 75% of the amount loaded).[18]

1.   Dissolve samples containing AEs in 10 to 50 μl of chloroform/methanol (9:1) and apply them 1.5 cm apart from the bottom of a silica gel G plate using a Hamilton syringe.

2.   Place the plate in a glass tank containing 50 to 100 ml of a solvent system composed of chloroform/methanol (9:1 or 95:5) or chloroform/methanol/ammonia (85:15:1). The latter system gives a better resolution between AEs and arachidonic acid (AA), which is delayed in its migration by the negatively charged carboxylate group. Glass tanks should be lined with filter paper to aid in saturating the chamber with solvent vapors.

3.   Allow the solvent to ascend about 14 cm up the plate; then remove the plate from the tank and evaporate the residual solvent in a fume hood.

4.   Spray the TLC plates with a 10% (w/v) solution of phosphomolybdic acid (Sigma) in ethanol and heat at 150°C for 5 min to visualize the lipids. AEs are localized by comparison with authentic standards applied in parallel lanes. Alternatively, visualization can be obtained by exposure to iodine vapor; the bands are then scraped off and dissolved in chloroform/methanol (1:1) for further analysis.

5.   The Rf of anandamide and other AEs under these conditions is approximately 0.5.

Note: *TLC solvents containing ammonia should not be used for the fractionation of oleamide, a lipid with anandamide-like properties.[19,20] The aldehyde octadecenal, present in tissues[21] or derived from plasmalogens[22] reacts rapidly with free ammonia to form oleamide, according to the following scheme:*

This reaction may occur substantially during a standard TLC run with a solvent system of chloroform/methanol/ammonia (D. Piomelli, unpublished results).

C. Analysis

Unambiguous identification of anandamide and allied AEs from biological samples can be obtained by GC/MS. The authors' laboratory uses a Hewlett-Packard 5890 GC equipped with an HP-5MS capillary column (30 m; internal diameter, 0.25 mm) and interfaced with a Hewlett-Packard 5972 MS. AEs are analyzed as trimethylsilylether (TMS) derivatives by electron-impact GC/MS. An isotope dilution assay for accurate quantitative measurements of AEs at very low concentrations is described below.

1.   Derivatization and GC/MS conditions

HPLC-purified AEs are converted to the corresponding TMS derivatives by treatment with 30 to 40 μl of *bis*(trimethylsilyl)trifluoroacetamide (BSTFA, Supelco) for 15 min at room temperature. This derivatization is recommended for its ease and high yields (>95%), as well as for the favorable GC/MS properties of the TMS derivatives of AEs.[4,17] The reaction products are dried under N_2, and the residue is dissolved in *n*-hexane and injected into the GC/MS in the splitless mode. Starting 1 min after the injection, the oven temperature is increased from 150 to 280°C at a rate of 8°C/min. The injector temperature is kept at 250°C and helium is used as the carrier gas. The ion energy used is 25 eV and the accelerating voltage 1.6 V.

2.   Mass spectral characteristics of acylethanolamides

Full mass spectra and characteristic fragmentation patterns for the TMS derivatives of three common AEs, anandamide (*cis*20:4), palmitylethanolamide (16:0), and oleylethanolamide (*cis*18:1) are shown in Figure 4. Fragments diagnostic for all compounds are found in the high mass range. These include molecular ions ([M]$^{+\cdot}$), as well as ions produced by the loss of one methyl group ([M-15]$^+$), one propyl group ([M-43]$^+$), and TMSOH group ([M-90]$^+$; the latter is absent from the spectrum of anandamide-TMS, which contains instead a fragment at m/z 328, [$^{-91}$M]$^+$), and m/z 175/179 (possibly corresponding to [H_2C=CO--NH--CH_2–CH_2–O–TMS]$^+$, which may be produced through McLafferty rearrangement). Similar fragment patterns are seen with other naturally occurring AEs. These include homo-γ-linolenylethanolamide ([M]$^{+\cdot}$, m/z 421), docosatetraenoylethanolamide ([M]$^{+\cdot}$, m/z 447), stearylethanolamide ([M]$^{+\cdot}$, m/z 399), linoleylethanolamide ([M]$^{+\cdot}$, m/z 395), and palmitoleylethanolamide ([M]$^{+\cdot}$, m/z 369).

3.   Isotope dilution assay

Isotope dilution analysis is a method of choice for the measurement of AEs because of their low concentrations in tissues, significant losses during isolation, and, most importantly, differential responses of the MS toward saturated and unsaturated molecules. The assay described below is based on stable isotope dilution followed by quantitative analysis by selected-ion monitoring (SIM) GC/MS.

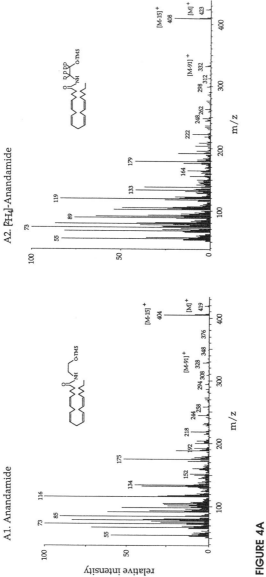

FIGURE 4A

Electron impact mass spectra of the TMS derivatives of anandamide (A1), [²H₄]anandamide (A2), palmitylethanolamide (B1), [²H₄]palmitylethanolamide (B2), oleylethanolamide (C1), and [²H₄]oleylethanolamide (C2).

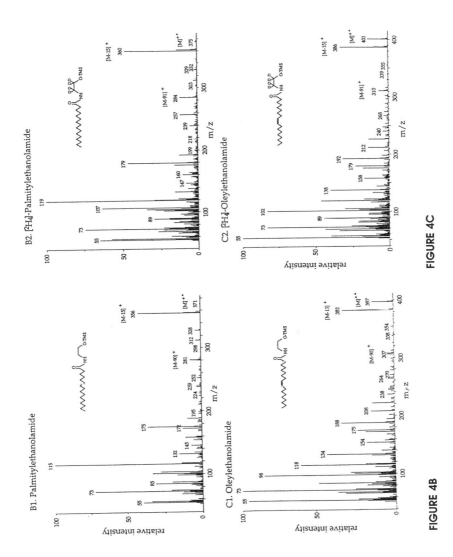

FIGURE 4B

FIGURE 4C

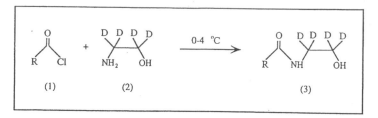

FIGURE 5

Synthesis of deuterated AE standards from fatty acyl chlorides and [^{2}H$_4$]ethanolamine in dichloromethane. Acyl chloride (1) and [^{2}H$_4$]ethanolamine (2) are allowed to react for 15 min at 0 to 4°C.

a. Procedure: Synthesis of Acylethanolamides and [^{2}H$_4$]Acylethanolamide Standards

Deuterated standards can be easily synthesized from the corresponding commercially available fatty acyl chlorides and [^{2}H$_4$]-labeled ethanolamine. This reaction, a scheme of which is shown in Figure 5, has been widely used for the preparation of unlabeled AEs.[1] It was selected because it allows one to prepare deuterated AE standards labeled on the ethanolamide moiety in a single step, which is both rapid (15 min) and efficient (average yield ≥95%). An alternative method involves the use of deuterated fatty acids as the starting material[23] and requires, therefore, two steps: (1) the conversion of the [^{2}H$_4$]fatty acid to its corresponding [^{2}H$_4$]acyl chloride and (2) the nucleophilic substitution of the chloride group with ethanolamine. The reaction time, as well as the losses and cost of this procedure, are significantly greater using this latter method, but the method may be used when fatty acyl chlorides are not commercially available. For GC/MS measurements of anandamide, a commercially available [^{2}H$_8$]arachido-nylethanolamide (Cayman Chemical, Ann Arbor, MI) can also be used as internal standard. The mass spectral properties of this standard are described in Reference 23.

1. Dissolve fatty acyl chlorides (Nu-Check Prep, Elysian, MN) in dichloromethane (10 mg/ml), mix with 2 equivalents of ethanolamine, and allow to react for 15 min at 0 to 4°C.

2. Stop reactions by adding 10 ml HPLC-grade water (Burdick & Jackson, Muskegon, MI). After vigorous mixing and phase separation, discard the upper aqueous phase to remove unreacted ethanolamine.

3. Wash the organic phases twice with water and concentrate to dryness in weighted glass vials. At constant weight, reconstitute the reaction products in methanol and store at –20°C for up to 3 months.

4. Determine the identity and chemical purity of the synthesized AEs and [^{2}H$_4$]AEs by TLC or GC/MS analysis.

b. Procedure: Construction of Calibration Curves

Standard calibration curves are constructed by adding a constant amount of deuterated standard (600 pmol) to increasing amounts (from 0 to 1000 pmol) of the corresponding unlabeled AE, followed by GC/MS analysis in the SIM mode. SIM peaks are then integrated, and the ratios of the areas under the peaks for the unlabeled AE and the [$^2H_4$]AE are plotted against the amounts of native AE injected into the GC/MS. The calibration curves for standard AEs are linear over the range of 0 to 1000 pmol ($r^2 = 0.99$) with a coefficient of variation of 4% at 2.5 pmol. Detection and quantification limits are in the high femtomole to low picomole range for all AEs. Representative calibration curves for anandamide, palmitylethanolamide, and oleylethanolamide are shown in Figure 6.

c. Procedure: Quantitative Analysis

To enhance recovery of the AEs and to allow for their GC/MS quantification, 600 pmol of [$^2H_4$]AE are added to the samples before extraction as internal standard and carrier. Identification and quantitation of endogenous AEs are achieved by monitoring in the SIM mode the AE fragments produced by the loss of one methyl group ([M-15]+). Figure 7 illustrates a representative SIM profile of anandamide from a rat blood plasma sample. The peak at m/z 404 corresponds to the [M-15]+ fragment of endogenous anandamide, while the peak at m/z 408 corresponds to the [M-15]+ ion of the [$^2H_4$]anandamide standard added before extraction. From the ratio of the areas under the peaks at m/z 404 and m/z 408, one can calculate the mass of anandamide in the sample by referring to the appropriate calibration curve.

Note: *Poor recoveries may result in overestimates of endogenous AE, because of unsatisfactory signal-to-noise ratio. This ratio should be at least 2:1 for reliable measurements.*

D.   Analysis of N-Acylphosphatidylethanolamines

Several lines of evidence indicate that AEs are produced by the cleavage of N-acyl PEs and that N-acyl PE biosynthesis is under the control of neural activity and second messengers.[4,5,24,25] This section describes some procedures for the extraction, purification, and analysis of N-acyl PEs from tissues.

1.   Extraction

Procedure

1.   To limit changes in N-acyl PEs levels, likely associated with post-mortem damage, immediately immerse the tissue in liquid N_2.
2.   Thaw frozen tissues and homogenize in methanol/Tris buffer (see Section II.A.2).
3.   Extract lipids twice in a mixture of chloroform, methanol, and buffer, adjusted to a volume ratio of 2:1:1.
4.   Subject N-acyl PE-containing fractions to chromatography on silica gel G columns as previously described, and elute N-acyl PEs with 2 ml of chloroform/methanol (6:4).

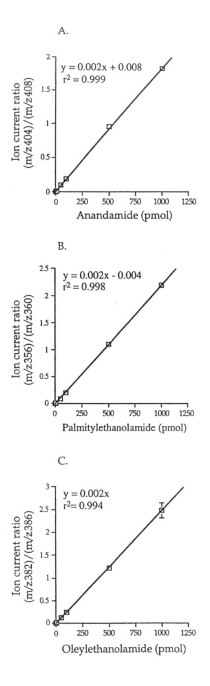

FIGURE 6

Representative standard curves for the isotope dilution quantification of anandamide (A), palmityletha-nolamide (B), and oleylethanolamide (C). Equations of the regression lines and regression coefficients (r^2) are indicated. Results represent the mean ± SEM of three independent measurements.

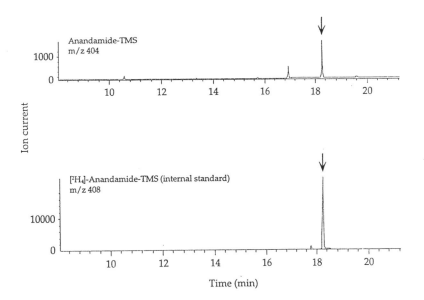

FIGURE 7
Representative GC chromatograms of a 2 ml rat blood plasma sample showing the presence of components with the chromatographic properties of anandamide. Anandamide was partially purified by normal-phase HPLC and analyzed by GC/MS in the SIM mode as TMS derivative as described in Section II.C.3c. The arrows indicate the retention times of authentic standards.

2.   Purification and identification

a. *Procedure: TLC*

N-acyl PE-containing fractions from column chromatography can be analyzed by one-dimensional TLC, using a solvent system of chloroform/methanol/ammonium hydroxide (80:20:1). Alternatively, *N*-acyl PEs can be analyzed by two-dimensional TLC or high-performance TLC; the latter is particularly useful for a more effective fractionation of small amounts of lipid material (generally, the amount of sample applied tends to be larger than that used for separations in a single direction). Figure 8 shows a fractionation of rat brain *N*-acyl PEs by two-dimensional TLC.

1. Apply samples as a spot at the lower right-hand corner of the plate, 1.5 cm apart from the edge.
2. Place the plate in a chromatographic chamber containing 50 to 100 ml of chloroform/ methanol/ammonium hydroxide (80:20:1). Allow the solvent front to migrate to 1.0 cm from the top of the plate. Remove the plate and dry at room temperature in a fume hood.
3. Rotate the plate 90° clockwise and place it in a chamber containing 50 to 100 ml of chloroform/methanol/acetic acid (80:20:1) for separation in the second dimension.
4. Remove the plate when the solvent front is 1.0 cm from the top edge. Dry the lipids and visualize them as described in Section II.B.3.

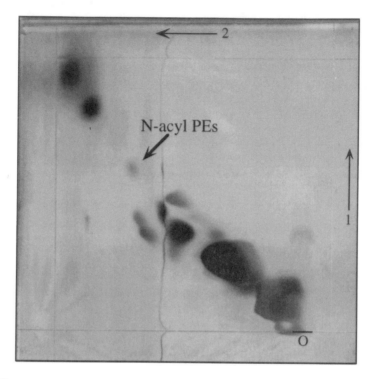

FIGURE 8

Identification of brain *N*-acyl PEs by two-dimension HPTLC. Phosphomolybdic acid staining revealed the presence of a lipid component (arrow) with the chromatographic properties of *N*-acyl PEs. (From Cadas, H. et al., *J. Neurosci.*, 17, 1226, 1997. With permission from the *Journal of Neuroscience*.)

b. Procedure: HPLC

N-acyl PE-containing fractions may be fractionated by reverse-phase HPLC on a µBondapak C18 column using a gradient of water in methanol (from 30 to 0% over 10 min) at a flow rate of 1.5 ml/min. This procedure resolves *N*-acyl PEs (retention time, about 14 min) from other lipid components that may be present in the *N*-acyl PE-containing fractions (e.g., PE, phosphatidylcholine, sphingomyelin, cerebrosides, diacylglycerols, cholesterol, etc.); however, *N*-acyl PEs differing in the *N*-acyl group are not resolved. The effluent fractions may be monitored through an evaporative light-scattering detector (for review see Reference 26) equipped with a photomultiplier. This technique was applied successfully to primary cultures of rat brain neurons[25] (Figure 9), but it was found difficult to utilize with samples containing larger amounts of lipid material (e.g., brain tissue).

c. Procedure: GC/MS

To confirm identification and determine the molecular composition of various *N*-acyl PEs, partially purified *N*-acyl PEs are digested with a bacterial phospholipase D (PLD).

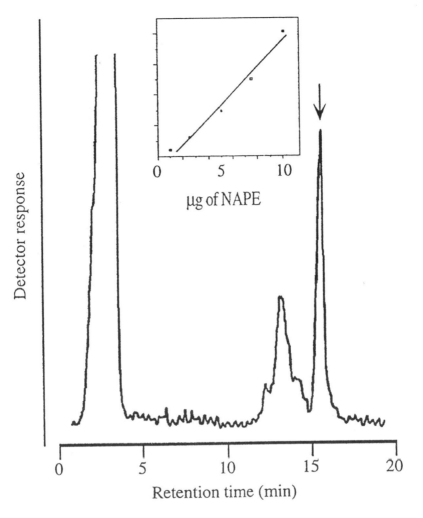

FIGURE 9
Quantitative analysis of neuronal *N*-acyl PEs by HPLC coupled to evaporative light-scattering detection.
The arrow indicates the retention time of synthetic *N*-arachidonyl PE. The response of the light-scattering
detector as a function of injected synthetic *N*-arachidonyl PE (NAPE) is illustrated in the inset. Photomultiplier voltage was set at 800 V, and nebulization temperature was set at 90°C.

The enzyme quantitatively hydrolyzes the distal phosphodiester bond of *N*-acyl PEs,
releasing the corresponding AEs (Figure 10).

1. Evaporate lipid fractions containing *N*-acyl PEs to dryness, reconstitute in ethylether
    (0.5 ml) and incubate in 40 mM MOPS buffer, pH 5.7, containing 250 U/ml *Strepto-
    myces chromofuscus* PLD (Sigma) and 15 m*M* CaCl$_2$ (2 ml). Incubations are carried
    out for 2 h at 37°C with shaking.

2. After addition of internal standards, extract the reaction products with chloroform/
    methanol (2:1), fractionate by normal-phase HPLC, and analyze by GC/MS as
    described in Sections II.B and C.

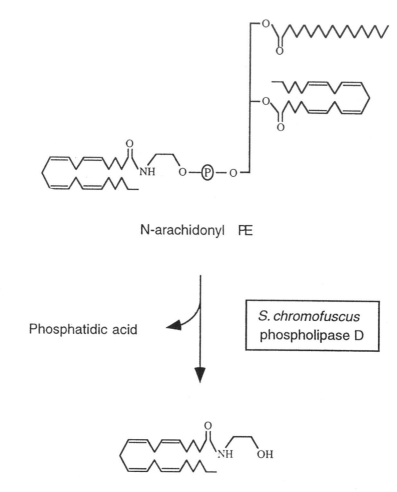

N-arachidonyl PE

Phosphatidic acid

S. chromofuscus
phospholipase D

anandamide

FIGURE 10

Identification of *N*-arachidonyl PE. Anandamide produced by the digestion of partially purified *N*-acyl PEs with *Streptomyces chromofuscus* PLD is subsequently purified by normal-phase HPLC, and analyzed by GC/MS.

Reagents Needed _____

Chemicals

Methanol (Burdick & Jackson, Muskegon, MI)
Chloroform (Burdick & Jackson)
Hexane (Burdick & Jackson)
Isopropyl alcohol (Mallinckrodt, Paris, KY)
Acetone (Burdick & Jackson) cooled to –20°C

bis(Trimethylsilyl)trifluoroacetamide (BSTFA) (Supelco, Bellefonte, PA)

Fatty acyl chlorides (Nu-Check Prep, Elysian, MN)

[²H₄]Ethanolamine (isotopic atom enrichment = 98%) (Cambridge Isotope Laboratories, Andover, MA)

All other chemicals from Sigma (St Louis, MO)

Buffers

Krebs Tris Buffer, pH 7.4

NaCl 136 mM

KCl 5 mM

MgCl₂ 1.2 mM

CaCl₂ 2.5 mM

Glucose 10 mM

Trizma base 20 mM

MOPS Buffer, pH 5.7

3-[N-Morpholino]propanesulfonic acid 40 mM

CaCl₂ 15 mM

III. Discussion

Extraction with chloroform/methanol and fractionation by column chromatography provide a rapid and inexpensive approach to separate tissue AEs from a variety of contaminants and represent two essential steps before HPLC and GC/MS analyses. Alternatively, AEs can be fractionated by TLC; this technique provides greater resolution than column chromatography, but it results in greater losses of sample and is not recommended for routine quantitative measurements. Unambiguous identification and quantification of AEs and their precursor N-acyl PEs may be achieved by GC/MS. The isotope dilution method described in this chapter has the accuracy and sensitivity necessary to measure AEs at trace concentrations, and it may be applied to determine fluctuations of AE levels in biological fluids under various physiological and pathological conditions.

Analysis of the AEs is hindered by their hydrophobic nature, their presence in tissues in very low amounts, and their tendency to adhere to biological membranes, glass, and plastic surfaces. In view of the expanding interest paid to the biology of these compounds, new approaches and more sensitive methodologies are needed. Solid-phase extraction, which consists of the adsorption of lipids onto a solid adsorbent (such as silica gel or octadecylsilane-bonded silica), may represent an alternative to the liquid–liquid extraction described here. Moreover, enhanced sensitivity and greater flexibility may be achieved by HPLC/MS.[23,27] This technique requires only relatively simple sample work-up and may be applied to the measurement of compounds, such as N-acyl PEs, which are not directly amenable to GC/MS analysis. Therefore, an increasing application of such technologies, which have had a limited use because of the high cost of the equipment, may be foreseen as these instruments become more readily available.

References

1. **Devane, W., Hanus, L., Breuer, A., Pertwee, R., Stevenson, L., Griffin, G., Gibson, D., Mandelbaum, D., Etinger, A., and Mechoulam, R.,** Isolation and structure of a brain constituent that binds to the cannabinoid receptor, *Science*, 258, 1946, 1992.

2. **Mechoulam, R., Hanus, L., and Martin, B.,** Search for endogenous ligands of the cannabinoid receptor, *Biochem. Pharmacol.*, 48, 1537, 1994.

3. **Di Marzo, V., Fontana, A., Cadas, H., Schinelli, S., Cimino, G., Schwartz, J.-C., and Piomelli, D.,** Formation and inactivation of endogenous cannabinoid anandamide in central neurons, *Nature*, 372, 686, 1994.

4. **Cadas, H., di Tomaso, E., and Piomelli, D.,** Occurrence and biosynthesis of endogenous cannabinoid precursor, N-arachidonoyl phosphatidylethanolamine, in rat brain, *J. Neurosci.*, 17, 1226, 1997.

5. **Sugiura, T., Kondo, S., Sukagawa, A., Tonegawa, T., Nakane, S., Yamashita, A., Ishima, Y., and Waku, K.,** Transacylase-mediated and phosphodiesterase-mediated synthesis of N-arachidonoylethanolamine, an endogenous cannabinoid-receptor ligand, in rat brain microsomes, *Eur. J. Biochem.*, 240, 53, 1996.

6. **Desarnaud, F., Cadas, H., and Piomelli, D.,** Anandamide amidohydrolase activity in rat brain microsomes, *J. Biol. Chem.*, 270, 6030, 1995.

7. **Ueda, N., Kurahashi, Y., Yamamoto, S., and Tokunaga, T.,** Partial purification and characterization of the porcine brain enzyme hydrolyzing and synthesizing anandamide, *J. Biol. Chem.*, 270, 23823, 1995.

8. **Cravatt, B. J., Giang, D. K., Mayfield, S. P., Boger, D. L., Lerner, R. A., and Gilula, N. B.,** Molecular characterization of an enzyme that degrades neuromodulatory fatty-acid amides, *Nature*, 384, 83, 1996.

9. **Beltramo, M., Stella, N., Calignano, A., Lin, S. Y., Makriyannis, A., and Piomelli, D.,** Functional role of high-affinity anandamide transport, as revealed by selective inhibition, *Science*, 277, 1094, 1997.

10. **Kurahashi, Y., Ueda, N., Suzuki, H., Suzuki, M., and Yamamoto, S.,** Reversible hydrolysis and synthesis of anandamide demonstrated by recombinant rat fatty-acid amide hydrolase, *Biochem. Biophys. Res. Commun.*, 237, 512, 1997.

11. **Schmid, H. H. O., Schmid, P. C., and Natarajan, V.,** The N-acylation-phosphodiesterase pathway and cell signalling, *Chem. Phys. Lipids*, 80, 133, 1996.

12. **Wagner, J. A., Varga, K., Ellis, E. F., Rzigalinski, B. A., Martin, B. R., and Kunos, G.,** Activation of peripheral CB_1 cannabinoid receptors in haemorrhagic shock, *Nature*, 390, 518, 1997.

13. Calignano, A., La Rana, G., Makriyannis, A., Lin, S. Y., Beltramo, M., and Piomelli, D., Inhibition of intestinal motility by anandamide, an endogenous cannabinoid, *Eur. J. Pharmacol.*, 340, R7, 1997.

14. Kaminski, N. E., Abood, M. E., Kessler, F. K., Martin, B. R., and Schatz, A. R., Identification of a functionally relevant cannabinoid receptor on mouse spleen cells that is involved in cannabinoid-mediated immune modulation, *Mol. Pharmacol.*, 42, 736, 1992.

15. Schmid, P. C., Krebsbach, R. J., Perry, S. R., Dettmer, T. M., Maasson, J. L., and Schmid, H. H. O., Occurrence and postmortem generation of anandamide and other long-chain N-acylethanolamines in mammalian brain, *FEBS Lett.*, 375, 117, 1995.

16. Kempe, K., Hsu, F., Bohrer, A., and Turk, J., Isotope dilution mass spectrometry measurements indicate that arachidoylethanolamide, the proposed endogenous ligand of the cannabinoid receptor, accumulates in rat brain tissue post mortem but is contained at low levels in or is absent from fresh tissue, *J. Biol. Chem.*, 271, 17287, 1996.

17. Giuffrida, A. and Piomelli, D., Isotope dilution GC/MS determination of anandamide and other fatty acylethanolamides in rat blood plasma, *FEBS Lett.*, 422, 373, 1998.

18. Fontana, A., Di Marzo, V., Cadas, H., and Piomelli, D., Analysis of anandamide, an endogenous cannabinoid substance, and of other natural *N*-acylethanolamines, *Prostag. Leukot. Ess. Fatty Acids*, 53, 301, 1995.

19. Cravatt, B. F., Prospero-Garcia, O., Siuzdak, G., Gilula, N. B., Heriksen, S. J., Boger, D. L., and Lerner, R. A., Chemical characterization of a family of brain lipids that induce sleep, *Science*, 268, 1506, 1995.

20. Bisogno, T., Sepe, N., De Petrocellis, L., Mechoulam, R., and Di Marzo, V., The sleep inducing factor oleamide is produced by mouse neuroblastoma cells, *Biochem. Biophys. Res. Commun.*, 239, 473, 1997.

21. Gilbertson, J. R., Johnson, R. C., Gelman, R. A., and Buffenmayer, C., Natural occurrence of free fatty aldehydes in bovine cardiac muscle, *J. Lipid Res.*, 13, 491, 1972.

22. Spener, F. and Mangold, H. K., Composition of alkoxylipids of human heart and aorta, *J. Lipid Res.*, 10, 609, 1969.

23. Felder, C. C., Nielsen, A., Briley, E. M., Palkovits, M., Priller, J., Axelrod, J., Nguyen, D. N., Richardson, J. M., Riggin, R. M., Koppel, G. A., Paul, S. M., and Becker, G. W., Isolation and measurement of the endogenous cannabinoid receptor agonist, anandamide, in brain and peripheral tissues of human and rat, *FEBS Lett.*, 393, 231, 1996.

24. Schmid, H. H. O., Schmid, P. C., and Natarajan, V., N-Acylated glycerophospholipids and their derivatives, *Prog. Lipid Res.*, 29, 1, 1990.

25. Cadas, H., Gaillet, S., Beltramo, M., Venance, L., and Piomelli, D., Biosynthesis of an endogenous cannabinoid precursor in neurons and its control by calcium and cAMP, *J. Neurosci.*, 16, 3934, 1996.

26. Christie, W., *HPLC and Lipids: A Practical Guide*, Pergamon, Oxford, 1987, 23.

27. Koga, D., Santa, T., Fukushima, T., Homma, H., and Imai, K., Liquid chromatographic-atmospheric pressure chemical ionization mass spectrometric determination of anandamide and its analogs in rat brain and peripheral tissues, *J. Chromatog. B*, 690, 7, 1997.

Index